Principles of Crop Production

Edited by
Kamden Carney

Larsen & Keller
www.larsen-keller.com

Principles of Crop Production
Edited by Kamden Carney
ISBN: 978-1-63549-080-0 (Hardback)

☐ Larsen & Keller

Published by Larsen and Keller Education,
5 Penn Plaza,
19th Floor,
New York, NY 10001, USA

Cataloging-in-Publication Data

Principles of crop production / edited by Kamden Carney.
        p. cm.
Includes bibliographical references and index.
ISBN 978-1-63549-080-0
1. Crop science. 2. Crops. 3. Agriculture. I. Carney, Kamden.
SB91 .P75 2017
633--dc23

The publisher's policy is to use permanent paper from mills that operate a sustainable forestry policy. Furthermore, the publisher ensures that the text paper and cover boards used have met acceptable environmental accreditation standards.

Printed and bound in the United States of America.

For more information regarding Larsen and Keller Education and its products, please visit the publisher's website www.larsen-keller.com

Table of Contents

Preface

This book deals with the concepts and methods related to crop production. It talks in detail about the various techniques used in this field. Crop production refers to that branch of agriculture, which deals with the growth and cultivation of crops keeping in mind their use as food or fiber. Different approaches, evaluations and methodologies on crop production have been included in this text. Various techniques of crop production along with technological progress that have implication are glanced at in this book. Some of the diverse topics covered in it address the varied branches that fall under this category. This textbook attempts to assist those with a goal of delving into the field of crop production.

A short introduction to every chapter is written below to provide an overview of the content of the book:

Chapter 1 - Crop yield is the measure of output or returns that is gained through agriculture in any area. The more surplus that is produced from this land, the more livestock can be maintained. The chapter on crop yield offers an insightful focus, keeping in mind the subject matter; **Chapter 2 -** Crops are plants that are harvested for food, for manufacture or for medical purposes. Some of the types of crops included in this chapter are cash crops, industrial crops, kharif crops, cover crops, fiber crops and rabi crops. This section helps the readers in developing an in depth understanding of crops; **Chapter 3 -** The processes of crop production are monoculture, intercropping, polyculture, organic farming, intensive farming, milpa, shifting cultivation etc. Monoculture is the practice of growing a single crop at a time whereas polyculture is the growing of more than one crop at one particular time. The section serves as a source to understand the major processes related to crop production; **Chapter 4 -** The essential aspects of agricultural productivity included in this chapter are sustainable agriculture, green revolution, water resource management, nutrient management, ecological sanitation etc. Sustainable agriculture is a sustainable way of farming, and helps in studying the relation between organisms and their environment. The chapter discusses the aspects of agricultural productivity in a critical manner providing key analysis on the topic; **Chapter 5 -** Fertilizers are materials that aid in the growth of plants. Fertilizers can be classified in several ways, such as single nutrient fertilizers, multinutrient fertilizers and micronutrients. Some of the features of agricultural productivity are pesticide, herbicide, liming, animal feed and irrigation. The chapter on fertilizers and pesticides offers an insightful focus, keeping in mind the topic; **Chapter 6 -** The environmental impacts of agriculture depends on the practices that are performed by the farmers of that particular region. The aspects discussed in this chapter are climate change and agriculture, genetically modified crops, environmental impact of irrigation, environmental impact of pesticides and plasticulture. The section serves as a source to understand the major impacts of agriculture; **Chapter 7 -** Technological innovation has played a very important role in the productivity of crops. Seeds and crops can be genetically altered to increase yield, growth rate as well as withstand disease and pests. This chapter elucidates the changes that have been made possible by technology in agriculture.

Finally, I would like to thank my fellow scholars who gave constructive feedback and my family members who supported me at every step.

Editor

Introduction to Crop Yield

Crop yield is the measure of output or returns that is gained through agriculture in any area. The more surplus that is produced from this land, the more livestock can be maintained. The chapter on crop yield offers an insightful focus, keeping in mind the subject matter.

In agriculture, crop yield (also known as "agricultural output") refers to both the measure of the yield of a crop per unit area of land cultivation, and the seed generation of the plant itself (e.g. if three grains are harvested for each grain seeded, the resulting yield is 1:3). The figure, 1:3 is considered by agronomists as the minimum required to sustain human life.

One of the three seeds must be set aside for the next planting season, the remaining two either consumed by the grower, or one for human consumption and the other for livestock feed. The higher the surplus, the more livestock can be established and maintained, thereby increasing the physical and economic well-being of the farmer and his family. This, in turn, resulted in better stamina, better over-all health, and better, more efficient work. In addition, the more the surplus the more draft animals — horse and oxen — could be supported and harnessed to work, and manure, the soil thereby easing the farmer's burden. Increased crop yields meant few hands were needed on farm, freeing them for industry and commerce. This, in turn, led to the formation and growth of cities.

Formation and growth of cities meant an increased demand for food stuffs by non-farmers, and their willingness to pay for it. This, in turn, led the farmer to (further) innovation, more intensive farming, the demand/creation of new and/or improved farming implements, and a quest for improved seed which improved crop yield. Thus allowing the farmer to raise his income by bringing more food to non-farming (city) markets.

Measurement

The unit by which the yield of a crop is measured is kilograms per hectare or bushels per acre.

History

Historically speaking, a major increase in crop yield took place in the early eighteenth century with the end of the ancient, wasteful cycle of the three-course system of crop rotation whereby a third of the land lay fallow every year and hence taken out of human food, and animal feed, production.

It was to be replaced by the four-course system of crop rotation, devised in England in 1730 by Charles Townshend, 2nd Viscount Townshend or "Turnip" Townshend during the British Agricultural Revolution, as he was called by early detractors.

In the *first year* wheat or oats were planted; in the *second year* barley or oats; in the *third year* clover, rye, rutabaga and/or kale were planted; in the *fourth year* turnips were planted but not harvested. Instead, sheep were driven on to the turnip fields to eat the crop, trample the leavings under their feet into the soil, and by doing all this, fertilize the land with their droppings. In the fifth year (or *first year* of the new rotation), the cycle began once more with a planting of wheat or oats, in an average, a thirty percent increased yield.

Crops and its Types

Crops are plants that are harvested for food, for manufacture or for medical purposes. Some of the types of crops included in this chapter are cash crops, industrial crops, kharif crops, cover crops, fiber crops and rabi crops. This section helps the readers in developing an in depth understanding of crops.

Crop

Domesticated plants

A crop is any cultivated plant, fungus, or alga that is harvested for food, clothing, livestock,fodder, biofuel, medicine, or other uses. In contrast, animals that are raised by humans are called livestock, except those that are kept as pets. Microbes, such as bacteria or viruses, are referred to as cultures. Microbes are not typically grown for food, but are rather used to alter food. For example, bacteria are used to ferment milk to produce yogurt.

Major crops include sugarcane, pumpkin, maize (corn), wheat, rice, cassava, soybeans, hay, potatoes and cotton.

Cash Crop

A cash crop is an agricultural crop which is grown for sale to return a profit. It is typically purchased by parties separate from a farm. The term is used to differentiate marketed crops from subsistence crops, which are those fed to the producer's own livestock or grown as food for the producer's family. In earlier times cash crops were usually only a small (but vital) part of a farm's total yield, while today, especially in developed countries, almost all crops are mainly grown for revenue. In the least developed countries, cash crops are usually crops which attract demand in more developed nations, and hence have some export value.

A cotton ball. Cotton is a significant cash crop. According to the National Cotton Council of America, in 2014, China was the world's largest cotton-producing country with an estimated 100,991,000 480-pound bales. India was ranked second at 42,185,000 480-pound bales.

Yerba mate (left, a key ingredient in the beverage known as mate), roasted by the fire, coffee beans (middle) and tea (right) are all used for caffeinated infusions and have cash crop histories.

Prices for major cash crops are set in commodity markets with global scope, with some local variation (termed as "basis") based on freight costs and local supply and demand balance. A consequence of this is that a nation, region, or individual producer relying on such a crop may suffer low prices should a bumper crop elsewhere lead to excess supply on the global markets. This system has been criticized by traditional farmers. Coffee is an example of a product that has been susceptible to significant commodity futures price variations.

Globalization

Issues involving subsidies and trade barriers on such crops have become controversial in discussions of globalization. Many developing countries take the position that the current international trade system is unfair because it has caused tariffs to be lowered in industrial goods while allowing for low tariffs and agricultural subsidies for agricultural goods.This makes it difficult for a developing nation to export its goods overseas, and forces developing nations to compete with imported goods which are exported from developed nations at artificially low prices. The practice of exporting at artificially low prices is known as dumping, and is illegal in most nations. Controversy over this issue led to the collapse of the Cancún trade talks in 2003, when the Group of 22 refused to consider agenda items proposed by the European Union unless the issue of agricultural subsidies was addressed.

Per Climate Zones

Arctic

The Arctic climate is generally not conducive for the cultivation of cash crops. However, one potential cash crop for the Arctic is *Rhodiola rosea*, a hardy plant used as a medicinal herb that grows in the Arctic. There is currently consumer demand for the plant, but the available supply is less than the demand (as of 2011).

Temperate

Cash crops grown in regions with a temperate climate include many cereals (wheat, rye, corn, barley, oats), oil-yielding crops (e.g. rapeseed, mustard seeds), vegetables (e.g. potatoes), tree fruit or top fruit (e.g. apples, cherries) and soft fruit (e.g. strawberries, raspberries).

A tea plantation in the Cameron Highlands in Malaysia

Subtropical

In regions with a subtropical climate, oil-yielding crops (e.g. soybeans) and some vegetables and herbs are the predominant cash crops.

Tropical

In regions with a tropical climate, coffee, cocoa, sugar cane, bananas, oranges, cotton and jute (a soft, shiny vegetable fiber that can be spun into coarse, strong threads), are common cash crops. The oil palm is a tropical palm tree, and the fruit from it is used to make palm oil.

By Continent and Country

Africa

Jatropha curcas is a cash crop used to produce biofuel.

Around 60 percent of African workers are employed in the agricultural sector, with about three-fifths of African farmers being subsistence farmers. For example, in Burkina Faso 85% of its residents (over two million people) are reliant upon cotton production for income, and over half of the country's population lives in poverty. Larger farms tend to grow cash crops such as coffee, tea, cotton, cocoa, fruit and rubber. These farms, typically operated by large corporations, cover tens of square kilometres and employ large numbers of laborers. Subsistence farms provide a source of food and a relatively small income for families, but generally fail to produce enough to make re-investment possible.

The situation in which African nations export crops while a significant amount of people on the continent struggle with hunger has been blamed on developed countries, including the United States, Japan and the European Union. These countries protect their own agricultural sectors, through high import tariffs and offer subsidies to their farmers, which some have contended is leading to the overproduction of commodities such as cotton, grain and milk. The result of this is that the global price of such products is continually reduced until Africans are unable to compete in world markets, except in cash crops that do not grow easily in temperate climates.

Africa has realized significant growth in biofuel plantations, many of which are on lands which were purchased by British companies. *Jatropha curcas* is a cash crop grown for biofuel production in Africa. Some have criticized the practice of raising non-food plants for export while Africa has problems with hunger and food shortages, and some studies have correlated the proliferation of land acquisitions, often for use to grow non-food cash crops with increasing hunger rates in Africa.

Australia

Australia produces significant amounts of lentils. It was estimated in 2010 that Australia would produce approximately 143,000 tons of lentils. Most of Australia's lentil harvest is exported to the Indian subcontinent and the Middle East.

United States

Oranges are a significant U.S. cash crop

Cash cropping in the United States rose to prominence after the baby boomer generation and the end of World War II. It was seen as a way to feed the large population boom and continues to be the main factor in having an affordable food supply in the United States. According to the 1997 U.S. Census of Agriculture, 90% of the farms in the United States are still owned by families, with an additional 6% owned by a partnership. Cash crop farmers have utilized precision agricultural technologies combined with time-tested practices to produce affordable food. Based upon United States Department of Agriculture (USDA) statistics for 2010, states with the highest fruit production quantities are California, Florida and Washington.

Various potato cultivars

Sliced sugarcane, a significant cash crop in Hawaii

Vietnam

Coconut is a cash crop of Vietnam.

Global Cash Crops

Coconut palms are cultivated in more than 80 countries of the world, with a total production of 61 million tonnes per year. The oil and milk derived from it are commonly used in cooking and frying; coconut oil is also widely used in soaps and cosmetics.

Sustainability of Cash Crops

Approximately 70% of the world's food is produced by 500 million smallholder farmers. For their livelihood they depend on the production of cash crops, basic commodities that are hard to differentiate in the market. The great majority (80%) of the world's farms measure 2 hectares or less. These smallholder farmers are mainly found in developing countries and are often unorganized, illiterate or enjoyed only basic education. Smallholder farmers have little bargaining power and incomes are low, leading to a situation in which they cannot invest much in upscaling their businesses. In general, farmers lack access to agricultural inputs and finance, and do not have enough knowledge on good agricultural and business practices. These high level problems are in many cases threatening the future of agricultural sectors and theories start evolving on how to secure a sustainable future for agriculture. Sustainable market transformations are initiated in which industry leaders work together in a pre-competitive environment to change market conditions. Sustainable intensification focuses on facilitating entrepreneurial farmers. To stimulate farm investment projects on access to finance for agriculture are also popping up. One example is the SCOPE methodology, an assessment tool that measures the management maturity and professionalism of producer organizations as to give financing organizations better insights in the risks involved in financing. Currently agricultural finance is always considered risky and avoided by financial institutions.

Black Market Cash Crops

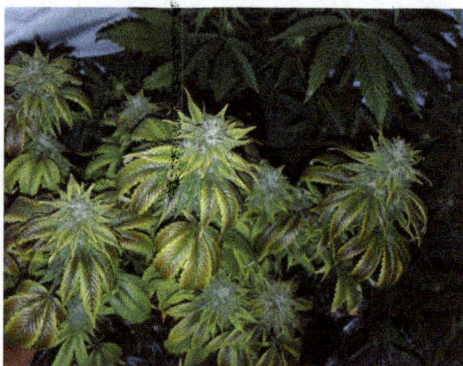

In the U.S., cannabis has been termed as a cash crop.

Coca, opium poppies and cannabis are significant black market cash crops, the prevalence of which varies. In the United States, cannabis is considered by some to be the most valuable cash crop. In 2006, it was reported in a study by Jon Gettman, a marijuana policy researcher, that in contrast to government figures for legal crops such as corn and wheat and using the study's projections for U.S. cannabis production at that time, cannabis was cited as "the top cash crop in 12 states and among the top three cash crops in 30." The study also estimated cannabis production at the time (in 2006) to be valued at $35.8 billion USD, which exceeded the combined value of corn at $23.3 billion and wheat at $7.5 billion.

Industrial Crop

An industrial crop, also called a non-food crop, is a crop grown to produce goods for manufacturing, for example of fibre for clothing, rather than food for consumption.

Purpose of Industrial Crops

Industrial crops is a designation given to an enterprise that attempts to raise farm sector income, and provide economic development activities for rural areas. Industrial crops also attempt to provide products that can be used as substitutes for imports from other nations.

Diversity

The range of crops with non-food uses is broad, but includes traditional arable crops like wheat, as well as less conventional crops like hemp and Miscanthus. Products made from non-food crops can be categorised by function:

Function	Products	Crop examples
Biofuels and bioenergy (energy crops)	Bioethanol, biobutanol, biodiesel, syngas and bioenergy	Algae, *Buchloe dactyloides*, *Jatropha*, and switchgrass
Building and construction	Hemp-lime building materials, Straw building materials, Insulation, Paints, varnishes	Hemp, wheat, linseed (flax), bamboo
Fiber	Paper, cloth, fabric, padding, string, twine, and rope	Coir, cotton, flax, hemp, manila hemp, papyrus, sisal
Pharmaceuticals (traditional) and therapeutic proteins (novel)	Drugs, botanical and herbal medicines, nutritional supplements, plant-made pharmaceuticals	Borage, *Cannabis sativa*, *Echinacea*, *Artemisia*, Tobacco
Renewable biopolymers	Plastics and packaging	Wheat, maize, potatoes
Speciality chemicals	Essential oils, printing ink, paper coatings	Lavender, oilseed rape, linseed, hemp

Kharif Crop

Kharif crops or monsoon crops are domesticated plants cultivated and harvested during the rainy (monsoon) season in the South Asia, which lasts between April and October depending on the area. Main kharif crops are millet and rice.

Kharif Season

Kharif crops are usually sown with the beginning of the first rains in July, during the south-west monsoon season. In Pakistan the kharif season starts on April 16th and lasts until October 15th. In India the kharif season varies by crop and state, with kharif starting at the earliest in May and ending at the latest in January, but is popularly considered to start in June and to end in October. Kharif stand in contrast with the Rabi crops, cultivated during the dry season. Both words came with the arrival of Mughals in the Indian subcontinent and are widely used ever-since. *Kharif* means "autumn" in Arabic. Since this period coincides with the beginning of autumn/winter in the Indian sub-continent, it is called "Kharif period".

Kharif crops are usually sown with the beginning of the first rains towards the end of May in the southern state of Kerala during the advent of south-west monsoon season. As the monsoon rains advance towards the north India, the sowing dates vary accordingly and reach July in north Indian states.

These crops are dependent on the quantity of rain water as well its timing. Too much, too little or at wrong time may lay waste the whole year's efforts.

Common Kharif Crops

- Rice (paddy and deepwater)
- Millet
- Maize (corn)

Catch Crop

White mustard grown as catch crop in Poland.

In agriculture, a catch crop is a fast-growing crop that is grown between successive plantings of a main crop.

For example, radishes that mature from seed in 25–30 days can be grown between rows of most vegetables, and harvested long before the main crop matures. Or, a catch crop can be planted between the spring harvest and fall planting of some crops.

Catch cropping is a type of succession planting. It makes more efficient use of growing space.

Catch crops are also crops that are sown to prevent minerals being flushed away from the soil. By using catch crops, such as grain (millet, ...) one can keep certain minerals not attached to the humous-clay connection (such as carbon (C) and other positively charged elements) in the soil for (many) years.

Cover Crop

A cover crop is a crop planted primarily to manage soil erosion, soil fertility, soil quality, water, weeds, pests, diseases, biodiversity and wildlife in an *agroecosystem* (Lu *et al.* 2000), an ecological system managed and largely shaped by humans across a range of intensities to produce food, feed, or fiber. Currently, not many countries are known for using the cover crop method.

Cover crops are of interest in sustainable agriculture as many of them improve the sustainability of agroecosystem attributes and may also indirectly improve qualities of neighboring natural ecosystems. Farmers choose to grow and manage specific cover crop types based on their own needs and goals, influenced by the biological, environmental, social, cultural, and economic factors of the food system in which they operate (Snapp *et al.* 2005). The farming practice of cover crops has been recognized as climate-smart agriculture by the White House.

Soil Erosion

Although cover crops can perform multiple functions in an agroecosystem simultaneously, they are often grown for the sole purpose of preventing soil erosion. Soil erosion is a process that can irreparably reduce the productive capacity of an agroecosystem. Dense cover crop stands physically slow down the velocity of rainfall before it contacts the soil surface, preventing soil splashing and erosive surface runoff (Romkens *et al.* 1990). Additionally, vast cover crop root networks help anchor the soil in place and increase soil porosity, creating suitable habitat networks for soil macrofauna (Tomlin *et al.* 1995). It keeps the enrichment of the soil good for the next few years.

Soil Fertility Management

One of the primary uses of cover crops is to increase soil fertility. These types of cover crops are referred to as "green manure." They are used to manage a range of soil macronutrients and micronutrients. Of the various nutrients, the impact that cover crops have on nitrogen management has received the most attention from researchers and farmers, because nitrogen is often the most limiting nutrient in crop production.

Often, green manure crops are grown for a specific period, and then plowed under before reaching full maturity in order to improve soil fertility and quality. Also the stalks left block the soil from being eroded.

Green manure crops are commonly leguminous, meaning they are part of the Fabaceae (pea) family. This family is unique in that all of the species in it set pods, such as bean, lentil, lupins and alfalfa. Leguminous cover crops are typically high in nitrogen and can often provide the required quantity of nitrogen for crop production. In conventional farming, this nitrogen is typically applied in chemical fertilizer form. This quality of cover crops is called fertilizer replacement value (Thiessen-Martens *et al.* 2005).

Another quality unique to leguminous cover crops is that they form symbiotic relationships with the rhizobial bacteria that reside in legume root nodules. Lupins is nodulated by the soil microorganism *Bradyrhizobium* sp. (Lupinus). Bradyrhizobia are encountered as microsymbionts in other leguminous crops (*Argyrolobium, Lotus, Ornithopus, Acacia, Lupinus*) of Mediterranean origin. These bacteria convert biologically unavailable atmospheric nitrogen gas (N 2) to biologically available ammonium (NH+ 4) through the process of biological nitrogen fixation.

Prior to the advent of the Haber-Bosch process, an energy-intensive method developed to carry out industrial nitrogen fixation and create chemical nitrogen fertilizer, most nitrogen introduced to ecosystems arose through biological nitrogen fixation (Galloway *et al.* 1995). Some scientists believe that widespread biological nitrogen fixation, achieved mainly through the use of cover crops, is the only alternative to industrial nitrogen fixation in the effort to maintain or increase future food production levels (Bohlool *et al.* 1992, Peoples and Craswell 1992, Giller and Cadisch 1995). Industrial nitrogen fixation has been criticized as an unsustainable source of nitrogen for food production due to its reliance on fossil fuel energy and the environmental impacts associated with chemical nitrogen fertilizer use in agriculture (Jensen and Hauggaard-Nielsen 2003). Such widespread environmental impacts include nitrogen fertilizer losses into waterways, which can lead to eutrophication (nutrient loading) and ensuing hypoxia (oxygen depletion) of large bodies of water.

An example of this lies in the Mississippi Valley Basin, where years of fertilizer nitrogen loading into the watershed from agricultural production have resulted in a hypoxic "dead zone" off the Gulf of Mexico the size of New Jersey (Rabalais *et al.* 2002). The ecological complexity of marine life in this zone has been diminishing as a consequence (CENR 2000).

As well as bringing nitrogen into agroecosystems through biological nitrogen fixation, types of cover crops known as "catch crops" are used to retain and recycle soil nitrogen already present. The catch crops take up surplus nitrogen remaining from fertilization of the previous crop, preventing it from being lost through leaching (Morgan *et al.* 1942), or gaseous denitrification or volatilization (Thorup-Kristensen *et al.* 2003).

Catch crops are typically fast-growing annual cereal species adapted to scavenge available nitrogen efficiently from the soil (Ditsch and Alley 1991). The nitrogen tied up in catch crop biomass is released back into the soil once the catch crop is incorporated as a green manure or otherwise begins to decompose.

An example of green manure use comes from Nigeria, where the cover crop *Mucuna pruriens* (velvet bean) has been found to increase the availability of phosphorus in soil after a farmer applies rock phosphate (Vanlauwe *et al.* 2000).

Soil Quality Management

Cover crops can also improve soil quality by increasing soil organic matter levels through the input of cover crop biomass over time. Increased soil organic matter enhances soil structure, as well as the water and nutrient holding and buffering capacity of soil (Patrick *et al.* 1957). It can also lead to increased soil carbon sequestration, which has been promoted as a strategy to help offset the rise in atmospheric carbon dioxide levels (Kuo *et al.* 1997, Sainju *et al.* 2002, Lal 2003).

Soil quality is managed to produce optimum circumstances for crops to flourish. The principal factors of soil quality are soil salination, pH, microorganism balance and the prevention of soil contamination.

Water Management

By reducing soil erosion, cover crops often also reduce both the rate and quantity of water that drains off the field, which would normally pose environmental risks to waterways and ecosystems downstream (Dabney *et al.* 2001). Cover crop biomass acts as a physical barrier between rainfall and the soil surface, allowing raindrops to steadily trickle down through the soil profile. Also, as stated above, cover crop root growth results in the formation of soil pores, which in addition to enhancing soil macrofauna habitat provides pathways for water to filter through the soil profile rather than draining off the field as surface flow. With increased water infiltration, the potential for soil water storage and the recharging of aquifers can be improved (Joyce *et al.* 2002).

Just before cover crops are killed (by such practices including mowing, tilling, discing, rolling, or herbicide application) they contain a large amount of moisture. When the cover crop is incorporated into the soil, or left on the soil surface, it often increases soil moisture. In agroecosystems where water for crop production is in short supply, cover crops can be used as a mulch to conserve water by shading and cooling the soil surface. This reduces evaporation of soil moisture. In other situations farmers try to dry the soil out as quickly as possible going into the planting season. Here prolonged soil moisture conservation can be problematic.

While cover crops can help to conserve water, in temperate regions (particularly in years with below average precipitation) they can draw down soil water supply in the spring, particularly if climatic growing conditions are good. In these cases, just before crop planting, farmers often face a tradeoff between the benefits of increased cover crop growth and the drawbacks of reduced soil moisture for cash crop production that season. C/N ratio is balanced with this application.

Weed Management

Cover crop in South Dakota

Thick cover crop stands often compete well with weeds during the cover crop growth period, and can prevent most germinated weed seeds from completing their life cycle and reproducing. If the cover crop is left on the soil surface rather than incorporated into the soil as a green manure after its growth is terminated, it can form a nearly impenetrable mat. This drastically reduces light transmittance to weed seeds, which in many cases reduces weed seed germination rates (Teasdale 1993). Furthermore, even when weed seeds germinate, they often run out of stored energy for growth before building the necessary structural capacity to break through the cover crop mulch layer. This is often termed the *cover crop smother effect* (Kobayashi *et al.* 2003).

Some cover crops suppress weeds both during growth and after death (Blackshaw *et al.* 2001). During growth these cover crops compete vigorously with weeds for available space, light, and nutrients, and after death they smother the next flush of weeds by forming a mulch layer on the soil surface. For example, Blackshaw *et al.* (2001) found that when using *Melilotus officinalis* (yellow sweetclover) as a cover crop in an improved fallow system (where a fallow period is intentionally improved by any number of different management practices, including the planting of cover crops), weed biomass only constituted between 1-12% of total standing biomass at the end of the cover crop growing season. Furthermore, after cover crop termination, the yellow sweetclover residues suppressed weeds to levels 75-97% lower than in fallow (no yellow sweetclover) systems .

In addition to competition-based or physical weed suppression, certain cover crops are known to suppress weeds through allelopathy (Creamer *et al.* 1996, Singh *et al.* 2003). This occurs when certain biochemical cover crop compounds are degraded that happen to be toxic to, or inhibit seed germination of, other plant species. Some well known examples of allelopathic cover crops are *Secale cereale* (rye), *Vicia villosa* (hairy vetch), *Trifolium pratense* (red clover), *Sorghum bicolor* (sorghum-sudangrass), and species in the Brassicaceae family, particularly mustards (Haramoto and Gallandt 2004). In one study, rye cover crop residues were found to have provided between 80% and 95% control of early season broadleaf weeds when used as a mulch during the production of different cash crops such as soybean, tobacco, corn, and sunflower (Nagabhushana *et al.* 2001).

In a recent study released by the Agricultural Research Service (ARS) scientists examined how rye seeding rates and planting patterns affected cover crop production. The results show that planting more pounds per acre of rye increased the cover crop's production as well as decreased the amount of weeds. The same was true when scientists tested seeding rates on legumes and oats; a higher density of seeds planted per acre decreased the amount of weeds and increased the yield of legume and oat production. The planting patterns, which consisted of either traditional rows or grid patterns, did not seem to make a significant impact on the cover crop's production or on the weed production in either cover crop. The ARS scientists concluded that increased seeding rates could be an effective method of weed control.

Disease Management

In the same way that allelopathic properties of cover crops can suppress weeds, they can also break disease cycles and reduce populations of bacterial and fungal diseases (Everts 2002), and parasitic nematodes (Potter *et al.* 1998, Vargas-Ayala *et al.* 2000). Species in the Brassicaceae family, such as mustards, have been widely shown to suppress fungal disease populations through the release of naturally occurring toxic chemicals during the degradation of glucosinolade compounds in their plant cell tissues (Lazzeri and Manici 2001).

Pest Management

Some cover crops are used as so-called "trap crops", to attract pests away from the crop of value and toward what the pest sees as a more favorable habitat (Shelton and Badenes-Perez 2006). Trap crop areas can be established within crops, within farms, or within landscapes. In many cases the trap crop is grown during the same season as the food crop being produced. The limited area occupied by these trap crops can be treated with a pesticide once pests are drawn to the trap in large enough numbers to reduce the pest populations. In some organic systems, farmers drive over the trap crop with a large vacuum-based implement to physically pull the pests off the plants and out of the field (Kuepper and Thomas 2002). This system has been recommended for use to help control the lygus bugs in organic strawberry production (Zalom *et al.* 2001). Another example of trap crops are nematode resistance White mustard (*Sinapis alba*) and Radish (*Raphanus sativus*). They can be grown after a main (cereal) crop and trap nematodes, for example the beet cyst nematode and Columbian root knot nematode. When grown, nematodes hatch and are attracted to the roots. After entering the roots they cannot reproduce in the root due to a hypersensitive resistance reaction of the plant. Hence the nematode population is greatly reduced, by 70-99%, depending on species and cultivation time.

Other cover crops are used to attract natural predators of pests by providing elements of their habitat. This is a form of biological control known as habitat augmentation, but achieved with the use of cover crops (Bugg and Waddington 1994). Findings on the relationship between cover crop presence and predator/pest population dynamics have been mixed, pointing toward the need for detailed information on specific cover crop types and management practices to best complement a given integrated pest management strategy. For example, the predator mite *Euseius tularensis* (Congdon) is known to help control the pest citrus thrips in Central California citrus orchards. Researchers found that the planting of several different leguminous cover crops (such as bell bean, woollypod vetch, New Zealand white clover, and Austrian winter pea) provided sufficient pollen as a feeding source to cause a seasonal increase in *E. tularensis* populations, which with good timing could potentially introduce enough predatory pressure to reduce pest populations of citrus thrips (Grafton-Cardwell *et al.* 1999).

Diversity and Wildlife

Although cover crops are normally used to serve one of the above discussed purposes, they often simultaneously improve farm habitat for wildlife. The use of cover crops adds at least one more dimension of plant diversity to a cash crop rotation. Since the cover crop is typically not a crop of value, its management is usually less intensive, providing a window of "soft" human influence on the farm. This relatively "hands-off" management, combined with the increased on-farm heterogeneity created by the establishment of cover crops, increases the likelihood that a more complex trophic structure will develop to support a higher level of wildlife diversity (Freemark and Kirk 2001).

In one study, researchers compared arthropod and songbird species composition and field use between conventionally and cover cropped cotton fields in the Southern United States. The cover cropped cotton fields were planted to clover, which was left to grow in between cotton rows throughout the early cotton growing season (stripcover cropping). During the migration and breeding season, they found that songbird densities were 7–20 times higher in the cotton fields with integrated clover cover crop than in the conventional cotton fields. Arthropod abundance and biomass was also higher in the clover cover cropped fields throughout much of the songbird breeding season, which was attributed to an increased supply of flower nectar from the clover. The clover cover crop enhanced songbird habitat by providing cover and nesting sites, and an increased food source from higher arthropod populations (Cederbaum *et al.* 2004).

Energy Crop

A Department for Environment, Food and Rural Affairs energy crops scheme plantation in the United Kingdom. Energy crops of this sort can be used in conventional power stations or specialised electricity generation units, reducing the amount of fossil fuel-derived carbon dioxide emissions.

An energy crop is a plant grown as a low-cost and low-maintenance harvest used to make biofuels, such as bioethanol, or combusted for its energy content to generate electricity or heat. Energy crops are generally categorized as woody or herbaceous plants; many of the latter are grasses of the family *Graminaceae*.

Commercial energy crops are typically densely planted, high-yielding crop species which are processed to bio-fuel and burnt to generate power. Woody crops such as willow or poplar are widely utilised, as well as temperate grasses such as *Miscanthus* and *Pennisetum purpureum* (both known as elephant grass). If carbohydrate content is desired for the production of biogas, whole-crops such as maize, Sudan grass, millet, white sweet clover and many others, can be made into silage and then converted into biogas.

Through genetic modification and application of biotechnology plants can be manipulated to create greater yields, reduce associated costs and require less water. High energy yield can also be realized with existing cultivars.

Types of Energy Crops

Solid Biomass

Energy generated by burning plants grown for the purpose, often after the dry matter is pelletized. Energy crops are used for firing power plants, either alone or co-fired with other fuels. Alternatively they may be used for heat or combined heat and power (CHP) production.

Elephant grass (Miscanthus sinensis) is an experimental energy crop

To cover the increasing requirements of woody biomass, short rotation coppice (SRC) were applied to agricultural sites. Within this cropping systems fast growing tree species like willows and poplars are planted in growing cycles of three to five years. The cultivation of this cultures is dependent on wet soil conditions and could be an alternative for moist field sieds. However, an influence on local water conditions could not be excluded. This indicates that an establishment should exclude the vicinity to vulnarable wetland ecosystems.

Gas Biomass (Methane)

Anaerobic digesters or biogas plants can be directly supplemented with energy crops once they have been ensiled into silage. The fastest growing sector of German biofarming has been in the area of "Renewable Energy Crops" on nearly 500,000 ha (1,200,000 acres) of land (2006). Energy crops can also be grown to boost gas yields where feedstocks have a low energy content, such as manures and spoiled grain. It is estimated that the energy yield presently of bioenergy crops converted via silage to methane is about 2 GWh/km^2 (1.8×10^{10} BTU/sq mi). Small mixed cropping enterprises with animals can use a portion of their acreage to grow and convert energy crops and sustain the entire farms energy requirements with about one fifth of the acreage. In Europe and especially Germany, however, this rapid growth has occurred only with substantial government support, as in the German bonus system for renewable energy. Similar developments of integrating crop farming and bioenergy production via silage-methane have been almost entirely overlooked in N. America, where political and structural issues and a huge continued push to centralize energy production has overshadowed positive developments.

Liquid Biomass

Biodiesel

Coconuts sun-dried in Kozhikode, Kerala for making copra, the dried meat, or kernel, of the coconut. Coconut oil extracted from it has made copra an important agricultural commodity for many coconut-producing countries. It also yields coconut cake which is mainly used as feed for livestock.

Pure biodiesel (B-100), made from soybeans

European production of biodiesel from energy crops has grown steadily in the last decade, principally focused on rapeseed used for oil and energy. Production of oil/biodiesel from rape covers more than 12,000 km² in Germany alone, and has doubled in the past 15 years. Typical yield of oil as pure biodiesel may be is 100,000 L/km² (68,000 US gal/sq mi; 57,000 imp gal/sq mi) or more, making biodiesel crops economically attractive, provided sustainable crop rotations exist that are nutrient-balanced and preventative of the spread of disease such as clubroot. Biodiesel yield of soybeans is significantly lower than that of rape.

Typical oil extractable by weight	
Crop	**Oil %**
copra	62
castor seed	50
sesame	50
groundnut kernel	42
jatropha	40
rapeseed	37
palm kernel	36
mustard seed	35
sunflower	32
palm fruit	20
soybean	14
cotton seed	13

Bioethanol

Energy crops for biobutanol are grasses. Two leading non-food crops for the production of cellulosic bioethanol are switchgrass and giant miscanthus. There has been a preoccupation with cellulosic bioethanol in America as the agricultural structure supporting biomethane is absent in many regions, with no credits or bonus system in place.Consequently, a lot of private money and investor hopes are being pinned on marketable and patentable innovations in enzyme hydrolysis and the like.

Bioethanol also refers to the technology of using principally corn (maize seed) to make ethanol directly through fermentation, a process that under certain field and process conditions can consume as much energy as is the energy value of the ethanol it produces, therefore being non-sustainable. New developments in converting grain stillage (referred to as distillers grain stillage or DGS) into biogas energy looks promising as a means to improve the poor energy ratio of this type of bioethanol process.

By Dedication

Dedicated energy crops are non-food energy crops as giant miscanthus, switchgrass, jatropha, fungi, and algae. Dedicated energy crops are promising cellulose sources that can be sustainably produced in many regions of the United States.

Additionally, the green waste byproducts of food and non-food energy crops can be used to produce various biofuels.

Fiber Crop

Fiber crops are field crops grown for their fibres, which are traditionally used to make paper, cloth, or rope. The fibers may be chemically modified, like in viscose (used to make rayon and cellophane). In recent years materials scientists have begun exploring further use of these fibers in composite materials.

Fiber crops are generally harvestable after a single growing season, as distinct from trees, which are typically grown for many years before being harvested for such materials as wood pulp fiber or lacebark. In specific circumstances, fiber crops can be superior to wood pulp fiber in terms of technical performance, environmental impact or cost.

There are a number of issues regarding the use of fiber crops to make pulp. One of these is seasonal availability. While trees can be harvested continuously, many field crops are harvested once during the year and must be stored such that the crop doesn't rot over a period of many months. Considering that many pulp mills require several thousand tonnes of fiber source per day, storage of the fiber source can be a major issue.

Botanically, the fibers harvested from many of these plants are bast fibers; the fibers come from the phloem tissue of the plant. The other fiber crop fibers are seed padding, leaf fiber, or other parts of the plant.

Fiber Sources

Ancient Sanskrit on Hemp based Paper. Hemp Fiber was commonly used in the production of paper from 200 BCE to the Late 1800's.

Before the industrialisation of the paper production the most common fibre source was recycled fibres from used textiles, called rags. The rags were from hemp, linen and cotton. A process for removing printing inks from recycled paper was invented by German jurist Justus Claproth in 1774. Today this method is called deinking. It was not until the introduction of wood pulp in 1843 that paper production was not dependent on recycled materials from ragpickers.

Fiber Crops

- Bast fibers (Stem-skin fibers)
    - Esparto, a fiber from a grass
    - Jute, widely used, it is the cheapest fiber after cotton
    - Flax, produces linen
    - Indian hemp, the Dogbane used by native Americans
    - Hemp, a soft, strong fiber, edible seeds
    - Hoopvine, also used for barrel hoops and baskets, edible leaves, medicine
    - Kenaf, the interior of the plant stem is used for its fiber. Edible leaves.
    - Linden Bast
    - Nettles
    - Ramie, a nettle, stronger than cotton or flax, makes "China grass cloth"
    - Papyrus, a pith fiber, akin to a bast fiber
- Leaf fibers
    - Abacá, a banana, producing "manila" rope from leaves
    - Sisal, often termed agave
    - Bowstring Hemp, an old use of a common decorative agave, also Sansevieria roxburghiana, Sansevieria hyacinthoides
    - Henequen, an agave. A useful fiber, but not as high quality as sisal
    - Phormium, "New Zealand Flax"
    - Yucca, an agave
- Seed fibers and fruit fibers
    - Coir, the fiber from the coconut husk
    - Cotton

- o Kapok

- o Milkweed, grown for the filament-like pappus in its seed pods

- o Luffa, a gourd which when mature produces a sponge-like mass of xylem, used to make loofa sponge

- Other fibers (Leaf, fruit, and other fibers)

  - o Bamboo fiber, a viscose fiber like rayon, technically a semi-synthetic fiber

Fiber Dimensions

Source of pulp	Fiber length, mm	Fiber diameter, μm
Softwood	3.1	30
Hardwood	1.0	16
Wheat straw	1.5	13
Rice straw	1.5	9
Esparto grass	1.1	10
Reed	1.5	13
Bagasse	1.7	20
Bamboo	2.7	14
Cotton	25.0	20

Permanent Crop

A permanent crop is one produced from plants which last for many seasons, rather than being replanted after each harvest.

Traditionally, "arable land" included any land suitable for the growing of crops, even if it was actually being used for the production of permanent crops such as grapes or peaches. Modern agriculture—particularly organizations such as the CIA and FAO—prefer the term of art permanent cropland to describe such "cultivable land" that is not being used for annually-harvested crops such as staple grains. In such usage, permanent cropland is a form of "agricultural land" that includes grasslands and shrublands used to grow grape vines or coffee; orchards used to grow fruit or olives; and forested plantations used to grow nuts or rubber. It does not include, however, tree farms intended to be used for wood or timber.

Rabi Crop

Wheat

Barley

Rabi crops or Rabi harvest are agricultural crops sown in winter and harvested in the spring in the South Asia. The term is derived from the Arabic word for "spring", which is used in the Indian subcontinent, where it is the spring harvest (also known as the "winter crop").

The rabi crops are sown around mid-November, after the monsoon rains are over, and harvesting begins in April/May. The crops are grown either with rainwater that has percolated into the ground, or with irrigation. A good rain in winter spoils the rabi crops but is good for kharif crops.

The major rabi crop in India is wheat, followed by barley, mustard, sesame and peas. Peas are harvested early, as they are ready early: Indian markets are flooded with green peas from January to March, peaking in February.

Many crops are cultivated in both kharif and rabi seasons. The agriculture crops produced in India are seasonal in nature and highly dependent on these two monsoons.

Examples of Rabi Crops:

Cereals

- wheat (*Triticum aestvium*)

- oat (*Avena sativa*)

- barley

Seed plants

- alfalfa (Lucerne, *Medicago sativa*)

- linseed

- sesame

- cumin (*Cuminum cyminum, L*)

- coriander (*Coriandrum sativum, L*)

- mustard (*Brassica juncea L.*)

- fennel (*Foeniculum vulgare*)

- fenugreek (*Trigonella foenumgraecum, L*)

- isabgol (*Plantago ovata*)

Vegetables

- pea

- chickpea (Gram, *Cicer arientinum*)

- onion (*Allium cepa, L.*)

- tomato (*Solanum lycopersicum, L*)

- potato (*Solanum tuberosum*)

Nurse Crop

Oats as nurse crop for alfalfa

In agriculture, a nurse crop is an annual crop used to assist in establishment of a perennial crop. The widest use of nurse crops is in the establishment of legumaceous plants such as alfalfa, clover, and trefoil. Occasionally nurse crops are used for establishment of perennial grasses.

Nurse crops reduce the incidence of weeds, prevent erosion, and prevent excessive sunlight from reaching tender seedlings. Often the nurse crop can be harvested for grain, straw, hay, or pasture. Oats are the most common nurse crop, though other annual grains are also. Nurse cropping of tall or dense-canopied plants, may protect more vulnerable species through shading or by providing a wind break.

References

- Cruz, Von Mark V.; Dierig, David A. (2014). Industrial Crops: Breeding for BioEnergy and Bioproducts. Springer. pp. 9 and passim. ISBN 978-1-4939-1447-0.

- Ara Kirakosyan; Peter B. Kaufman (2009-08-15). Recent Advances in Plant Biotechnology. p. 169. ISBN 9781441901934. Retrieved 14 February 2013.

- Göttsching, Lothar; Pakarinen, Heikki (2000), "1", Recycled Fiber and Deinking, Papermaking Science and Technology, 7, Finland: Fapet Oy, pp. 12–14, ISBN 952-5216-07-1

- J. A. R. Lockhart; A. J. L. Wiseman (17 May 2014). Introduction to crop husbandry including grassland). Elsevier. p. 111. Retrieved 24 June 2015.

- Food and Agriculture Organization of the United Nations, Statistics Division (2009). "Maize, rice and wheat : area harvested, production quantity, yield". Archived from the original on January 14, 2013.

- Borders, Max; Burnett, H. Sterling (March 24, 2006). "Farm Subsidies: Devastating the World's Poor and the Environment". National Center for Policy Analysis. Retrieved April 6, 2012.

Techniques of Crop Production

The processes of crop production are monoculture, intercropping, polyculture, organic farming, intensive farming, milpa, shifting cultivation etc. Monoculture is the practice of growing a single crop at a time whereas polyculture is the growing of more than one crop at one particular time. The section serves as a source to understand the major processes related to crop production.

Monoculture

Monoculture is the agricultural practice of producing or growing a single crop, plant, or livestock species, variety, or breed in a field or farming system at a time. Polyculture, where more than one crop is grown in the same space at the same time, is the alternative to monoculture. Monoculture is widely used in both industrial farming and organic farming and has allowed increased efficiency in planting and harvest.

Continuous monoculture, or monocropping, where the same species is grown year after year, can lead to the quicker buildup of pests and diseases, and then rapid spread where a uniform crop is susceptible to a pathogen. The practice has been criticized for its environmental effects and for putting the food supply chain at risk. Diversity can be added both in time, as with a crop rotation or sequence, or in space, with a polyculture.

Diversity of crops in space and time; monocultures and polycultures, and rotations of both.					
			Diversity in time		
		Higher			
	Low	Cyclic	Dynamic (non-cyclic)		
Diversity in space	Low	Monoculture, one species in a field	Continuous monoculture, monocropping	Crop rotation (rotation of monocultures)	Sequence of monocultures
	Higher	Polyculture, two or more species intermingled in a field	Continuous polyculture	Rotation of polycultures	Sequence of polycultures

Oligoculture has been suggested to describe a crop rotation of just a few crops, as is practiced by several regions of the world.

The term monoculture is frequently applied for other uses to describe any group dominated by a single variety, e.g. social Monoculturalism, or in the field of musicology to describe the dominance of the American and British music-industries in Western pop music, or in the field of computer science to describe a group of computers all running identical software.

A monocultivated potato field

Land Use

The term is used in agriculture and describes the practice of planting the same cultivar over an extended area. Each cultivar has the same standardized planting, maintenance and harvesting requirements resulting in greater yields and lower costs.

It also is beneficial because a crop can be tailor-planted for a location that has special problems – like soil salt or drought or a short growing season.

When a crop is matched to its well-managed environment, a monoculture can produce higher yields than a polyculture. In the last 40 years, modern practices such as monoculture planting and the use of synthesized fertilizers have reduced the amount of additional land needed to produce food. However, planting the same crop in the same place each year depletes the nutrients from the earth that the plant relies on and leaves soil weak and unable to support healthy plant growth. Because soil structure and quality is so poor, farmers are forced to use chemical fertilizers to encourage plant growth and fruit production. These fertilizers, in turn, disrupt the natural makeup of the soil and contribute further to nutrient depletion. Monocropping also creates the spread of pests and diseases, which have to be treated with yet more chemicals. The effects of monocropping on the environment are severe when pesticides and fertilizers make their way into ground water or become airborne, creating pollution.

Forestry

In forestry, monoculture refers to the planting of one species of tree. Monoculture plantings provide great yields and more efficient harvesting than natural stands of trees. Single-species stands of trees are often the natural way trees grow, but the stands show a diversity in tree sizes, with dead trees mixed with mature and young trees. In forestry, monoculture stands that are planted and harvested as a unit provide limited resources for wildlife that depend on dead trees and openings, since all the trees are the same size; they are most often harvested by clearcutting, which drastically alters the habitat. The mechanical harvesting of trees can compact soils, which can adversely affect understory growth. Single-species planting of trees also are more vulnerable when infected with a pathogen, or are attacked by insects, and by adverse environmental conditions.

Lawns and Animals

Examples of monoculture include lawns and most field cs wheat or corn. The term is also used where a single breed of farm animal is raised in large-scale concentrated animal feeding operations (CAFOs). In the United States, The Livestock Conservancy was formed to protect nearly 200 endangered livestock breeds from going extinct, largely due to the increased reliance on just a handful of highly specialized breeds.

Disease

Crops used in agriculture are usually single strains that have been bred for high yield and resistant to certain common diseases. Since all plants in a monoculture are genetically similar, if a disease strikes to which they have no resistance, it can destroy entire populations of crops. Polyculture, which is the mixing of different crops, reduces the likelihood that one or more of the crops will be resistant to any particular pathogen. Studies have shown planting a mixture of crop strains in the same field to be effective at combating disease. Ending monocultures grown under disease conditions by introducing crop diversity has greatly increased yields. In one study in China, the planting of several varieties of rice in the same field increased yields of non-resistant strains by 89% compared to non-resistant strains grown in monoculture, largely because of a dramatic (94%) decrease in the incidence of disease, making pesticides less necessary. There is currently a great deal of international worry about the wheat leaf rust fungus, that has already decimated wheat crops in Uganda and Kenya, and is starting to make inroads into Asia as well. As much of the world's wheat crops are very genetically similar following the Green Revolution, the impacts of such diseases threaten agricultural production worldwide.

In Ireland, exclusive use of one variety of potato, the "lumper", led to the great famine. It was inexpensive food to feed the masses. Potatoes were propagated vegetatively with little to no genetic variation. When *Phytophthora infestans* arrived from the Americas in 1845 to Ireland, the lumper had no resistance to the disease leading to the nearly complete failure of the potato crop across Ireland. Had the farmers used multiple varieties of potato, the famine may not have occurred. Andean natives were cultivating three thousand varieties before the Spaniards arrived.

Many of today's livestock production systems rely on just a handful of highly specialized breeds. By focusing heavily on a single trait (output), other traits like fertility, disease resistance, vigor, and mothering instincts are sacrificed. In the early 1990s a few Holstein calves were observed to grow poorly and died in the first 6 months of life. They were all found to be homozygous for a mutation in the gene that caused Bovine Leukocyte Adhesion Deficiency. This mutation was found at a high frequency in Holstein populations world-wide. (15% among bulls in the US, 10% in Germany, and 16% in Japan.) By studying the pedigrees of affected and carrier animals the source of the mutation was tracked to a single bull that was widely used in the industry. Note that in 1990 there were approximately 4 million Holteins in the US making the affected population around 600,000 animals.

Polyculture

The environmental movement seeks to change popular culture by redefining the "perfect lawn" to be something other than a turf monoculture, and seeks agricultural policy that provides greater encouragement for more diverse cropping systems. Local food systems may also encourage growing multiple species and a wide variety of crops at the same time and same place. Heirloom gardening and raising heritage livestock breeds have come about largely as a reaction against monocultures in agriculture.

Polyculture

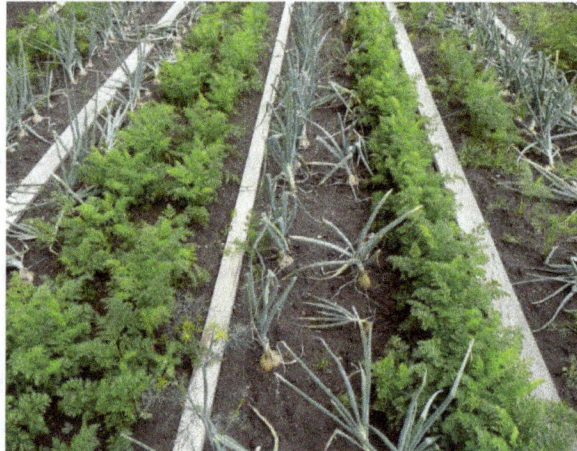

Polyculture providing useful within-field diversity: companion planting of carrots and onions. The onion smell puts off carrot root fly, while the smell of carrots puts off onion fly.

Polyculture is agriculture using multiple crops in the same space, providing crop diversity in imitation of the diversity of natural ecosystems, and avoiding large stands of single crops, or monoculture. It includes multi-cropping, intercropping, companion planting, beneficial weeds, and alley cropping. It is the raising at the same time and place of more than one species of plant or animal. Polyculture is one of the principles of permaculture.

Advantages

Polyculture, though it often requires more labor, has two main advantages over monoculture.

Polyculture reduces susceptibility to disease. For example, a study in China showed that planting several varieties of rice in the same field increased yields by 89%, largely because of a dramatic (94%) decrease in the incidence of disease, which made pesticides redundant.

Polyculture increases local biodiversity. This is one example of reconciliation ecology, or accommodating biodiversity within human landscapes. This may also form part of a biological pest control program

Multiple Cropping

In agriculture, multiple cropping is the practice of growing two or more crops in the same piece of land during a single growing season. It is a form of polyculture. It can take the form of double-cropping, in which a second crop is planted after the first has been harvested, or relay cropping, in which the second crop is started amidst the first crop before it has been harvested. A related practice, companion planting, is sometimes used in gardening and intensive cultivation of vegetables and fruits. One example of multi-cropping is tomatoes + onions + marigold; the marigolds repel some tomato pests.

Multiple cropping is found in many agricultural traditions. In the Garhwal Himalaya of India, a practice called baranaja involves sowing 12 or more crops on the same plot, including various types of beans, grains, and millets, and harvesting them at different times.

In the cultivation of rice, multiple cropping requires effective irrigation, especially in areas with a dry season. Rain that falls during the wet season permits the cultivation of rice during that period, but during the other half of the year, water cannot be channeled into the rice fields without an irrigation system. The Green Revolution in Asia led to the development of high-yield varieties of rice, which required a substantially shorter growing season of 100 days, as opposed to traditional varieties, which needed 150 to 180 days. Due to this, multiple cropping became more prevalent in Asian countries.

Intercropping

Intercropping is a multiple cropping practice involving growing two or more crops in proximity. The most common goal of intercropping is to produce a greater yield on a given piece of land by making use of resources that would otherwise not be utilized by a single crop. Careful planning is required, taking into account the soil, climate, crops, and varieties. It is particularly important not to have crops competing with each other for physical space, nutrients, water, or sunlight. Examples of intercropping strategies are planting a deep-rooted crop with a shallow-rooted crop, or planting a tall crop with a shorter crop that requires partial shade. Inga alley cropping has been proposed as an alternative to the ecological destruction of slash-and-burn farming.

When crops are carefully selected, other agronomic benefits are also achieved. Lodging-prone plants, those that are prone to tip over in wind or heavy rain, may be given structural support by their companion crop. Creepers can also benefit from structural support. Some plants are used to suppress weeds or provide nutrients. Delicate or light-sensitive plants may be given shade or protection, or otherwise wasted space can be utilized. An example is the tropical multi-tier system where coconut occupies the upper tier, banana the middle tier, and pineapple, ginger, or leguminous fodder, medicinal or aromatic plants occupy the lowest tier.

Intercropping of compatible plants also encourages biodiversity, by providing a habitat for a variety of insects and soil organisms that would not be present in a single-crop environment. This in turn can help limit outbreaks of crop pests by increasing predator biodiversity. Additionally, reducing the homogeneity of the crop increases the barriers against biological dispersal of pest organisms through the crop.

The degree of spatial and temporal overlap in the two crops can vary somewhat, but both requirements must be met for a cropping system to be an intercrop. Numerous types of intercropping, all of which vary the temporal and spatial mixture to some degree, have been identified. These are some of the more significant types:

- Mixed intercropping, as the name implies, is the most basic form in which the component crops are totally mixed in the available space.

- Row cropping involves the component crops arranged in alternate rows. Variations include alley cropping, where crops are grown in between rows of trees, and strip cropping, where multiple rows, or a strip, of one crop are alternated with multiple rows of another crop. A new version of this is to intercrop rows of solar photovoltaic modules with agriculture crops. This practice is called agrivoltaics.

- Temporal intercropping uses the practice of sowing a fast-growing crop with a slow-growing crop, so that the fast-growing crop is harvested before the slow-growing crop starts to mature.

- Further temporal separation is found in relay cropping, where the second crop is sown during the growth, often near the onset of reproductive development or fruiting, of the first crop, so that the first crop is harvested to make room for the full development of the second.

| Chili pepper intercropped with coffee in Colombia's southwestern Cauca Department | Coconut and *Tagetes erecta*, a multilayer cropping in India |

Companion Planting

Companion planting in gardening and agriculture is the planting of different crops in proximity for pest control, pollination, providing habitat for beneficial creatures, maximizing use of space, and to otherwise increase crop productivity. Companion planting is a form of polyculture.

Companion planting is used by farmers and gardeners in both industrialized and developing countries for many reasons. Many of the modern principles of companion planting were present many centuries ago in cottage gardens in England and forest gardens in Asia, and thousands of years ago in Mesoamerica.

History

In China, mosquito ferns (*Azolla* spp.) have been used for at least a thousand years as companion plants for rice crops. They host a cyanobacterium that fixes nitrogen from the atmosphere, and they block light from plants that would compete with the rice.

Companion planting was practiced in various forms by the indigenous peoples of the Americas prior to the arrival of Europeans. These peoples domesticated squash 8,000 to 10,000 years ago, then maize, then common beans, forming the Three Sisters agricultural technique. The cornstalk served as a trellis for the beans to climb, and the beans fixed nitrogen, benefitting the maize.

Companion planting was widely promoted in the 1970s as part of the organic gardening movement.It was encouraged for pragmatic reasons, such as natural trellising, but mainly with the idea that different species of plant may thrive more when close together.It is also a technique frequently used in permaculture, together with mulching, polyculture, and changing of crops.

Examples of Companion Plants

Nasturtium (*Tropaeolum majus*) is a food plant of some caterpillars which feed primarily on members of the cabbage family (brassicas), and some gardeners claim that planting them around brassicas protects the food crops from damage, as eggs of the pests are preferentially laid on the nasturtium.This practice is called trap cropping (using alternative plants to attract pests away from a main crop). However, while many trap crops have successfully diverted pests off of focal crops in small scale greenhouse, garden and field experiments, only a small portion of these plants have been shown to reduce pest damage at larger commercial scales.

The smell of the foliage of marigolds is claimed to deter aphids from feeding on neighbouring crops. Marigolds with simple flowers also attract nectar-feeding adult hoverflies, the larvae of which are predators of aphids.

Various legume crops benefit from being commingled with a grassy nurse crop. For example, common vetch or hairy vetch is planted together with rye or winter wheat to make a good cover crop or green manure (or both).

The terms "undersowing" and "overseeding" both involve intercropping as a type of companion planting. "Undersowing" conveys the idea of sowing the second crop among the young plants of the first crop (or in between the rows, if rows are used). A connotation of understory growth is conveyed, albeit exaggerated (because the first crop is not yet a dense canopy). "Overseeding" conveys the idea of broadcasting the seeds of the second crop over the existing first crop. This is analogous to overseeding a lawn to improve the mix of grasses present.

Versions

There are a number of systems and ideas using companion planting.

Square foot gardening attempts to protect plants from many normal gardening problems by packing them as closely together as possible, which is facilitated by using companion plants, which can be closer together than normal.

Another system using companion planting is the forest garden, where companion plants are intermingled to create an actual ecosystem, emulating the interaction of up to seven levels of plants in a forest or woodland.

Organic gardening may make use of companion planting, since many synthetic means of fertilizing, weed reduction and pest control are forbidden.

Host-finding Disruption

Recent studies on host-plant finding have shown that flying pests are far less successful if their host-plants are surrounded by any other plant or even "decoy-plants" made of green plastic, cardboard, or any other green material.

The host-plant finding process occurs in phases:

- The first phase is stimulation by odours characteristic to the host-plant. This induces the insect to try to land on the plant it seeks. But insects avoid landing on brown (bare) soil. So if only the host-plant is present, the insects will quasi-systematically find it by simply landing on the only green thing around. This is called (from the point of view of the insect) "appropriate landing". When it does an "inappropriate landing", it flies off to any other nearby patch of green. It eventually leaves the area if there are too many 'inappropriate' landings.

- The second phase of host-plant finding is for the insect to make short flights from leaf to leaf to assess the plant's overall suitability. The number of leaf-to-leaf flights varies according to the insect species and to the host-plant stimulus received from each leaf. The insect must accumulate sufficient stimuli from the host-plant to lay eggs; so it must make a certain number of consecutive 'appropriate' landings. Hence if it makes an 'inappropriate landing', the assessment of that plant is negative, and the insect must start the process anew.

Thus it was shown that clover used as a ground cover had the same disruptive effect on eight pest species from four different insect orders. An experiment showed that 36% of cabbage root flies laid eggs beside cabbages growing in bare soil (which resulted in no crop), compared to only 7% beside cabbages growing in clover (which allowed a good crop). Simple decoys made of green cardboard also disrupted appropriate landings just as well as did the live ground cover.

Companion Plant Categories

The use of companion planting can be of benefit to the grower in a number of different ways, including:

- Hedged investment – the growing of different crops in the same space increases the odds of some yield being given, even if one crop fails.

- Increased level interaction – when crops are grown on different levels in the same space, such as providing ground cover or one crop working as a trellis for another, the overall yield of a plot may be increased.

- Protective shelter is when one type of plant may serve as a wind break or provide shade for another.

- Pest suppression – some companion plants may help prevent pest insects or pathogenic fungi from damaging the crop, through chemical means.

- Predator recruitment and positive hosting – The use of companion plants that produce copious nectar or pollen in a vegetable garden (insectary plants) may help encourage higher populations of beneficial insects that control pests, as some beneficial predatory insects only consume pests in their larval form and are nectar or pollen feeders in their adult form.

- Trap cropping – some companion plants are claimed to attract pests away from others.

- Pattern disruption – in a monoculture pests spread easily from one crop plant to the next, whereas such easy progress may be disrupted by surrounding companion plants of a different type.

Three Sisters (Agriculture)

Three Sisters as featured on the reverse of the 2009 Native American U.S. dollar coin

The Three Sisters are the three main agricultural crops of various Native American groups in North America: winter squash, maize (corn), and climbing beans (typically tepary beans or common beans). The Iroquois, among others, used these "Three Sisters" as trade goods.

In a technique known as companion planting, the three crops are planted close together. Flat-topped mounds of soil are built for each cluster of crops. Each mound is about 30 cm (12 in) high and 50 cm (20 in) wide, and several maize seeds are planted close together in the center of each mound. In parts of the Atlantic Northeast, rotten fish or eels are buried in the mound with the maize seeds, to act as additional fertilizer where the soil is poor. When the maize is 15 cm (6 inches) tall, beans and squash are planted around the maize, alternating between the two kinds of seeds. The process to develop this agricultural knowledge took place over 5,000–6,500 years. Squash was domesticated first, with maize second and then beans being domesticated. Squash was first domesticated 8,000–10,000 years ago.

The three crops benefit from each other. The maize provides a structure for the beans to climb, eliminating the need for poles. The beans provide the nitrogen to the soil that the other plants use, and the squash spreads along the ground, blocking the sunlight, helping prevent the establishment of weeds. The squash leaves also act as a "living mulch", creating a microclimate to retain moisture in the soil, and the prickly hairs of the vine deter pests. Corn, beans, and squash contain complex carbohydrates, essential fatty acids and all eight essential amino acids, allowing most Native American tribes to thrive on a plant-based diet.

Native Americans throughout North America are known for growing variations of Three Sisters gardens. The milpas of Mesoamerica are farms or gardens that employ companion planting on a larger scale. The Anasazi are known for adopting this garden design in a drier environment. The Tewa and other Southwestern United States tribes often included a "fourth sister" known as "Rocky Mountain bee plant" (*Cleome serrulata*), which attracts bees to help pollinate the beans and squash.

The Three Sisters planting method is featured on the reverse of the 2009 US Sacagawea Native American dollar coin.

Beneficial Weed

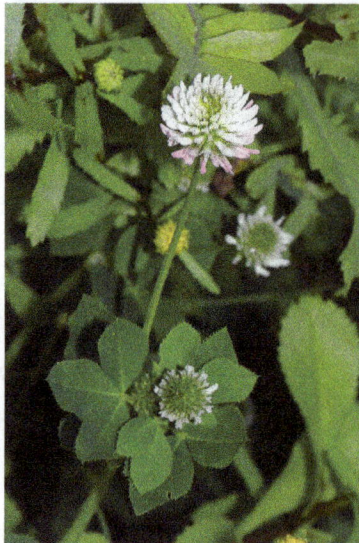

Clover was once included in grass seed mixes, because it is a legume that fertilizes the soil

A beneficial weed is a plant not generally considered domesticated and often viewed as a weed but which has some companion plant effect, is edible, contributes to soil health, or is otherwise beneficial. Beneficial weeds include many wildflowers, as well as other weeds that are commonly removed or poisoned.

Soil Health

Dandelions benefit neighboring plant health by bringing up nutrients and moisture with its deep tap root

Although erroneously assumed to compete with neighboring plants for food and moisture, some "weeds" provide the soil with nutrients, either directly or indirectly.

- For example, legumes, such as white clover, if they are colonized by the right bacteria (Rhizobium most often) add nitrogen to the soil through the process of nitrogen fixation, where the bacteria has a symbiotic relationship with its hosts roots, "fixing" atmospheric nitrogen (combining it with oxygen or hydrogen) making the nitrogen plant-available (NH_4 or NO_3).

- Others use deep tap roots to bring up nutrients and moisture from beyond the range of normal plants so that the soil improves in quality over generations of that plant's presence.

- Weeds with strong, widespread roots also introduce organic matter to the earth in the form of those roots, turning hard, dense clay dirt into richer, more fertile soil.

- Some plants like tomatoes and corn will "piggyback" on nearby weeds, allowing their relatively weak root systems to go deeper.

Pest Prevention

Crow garlic, like any allium, masks scents from pest insects, protecting neighboring plants

Many weeds protect nearby plants from insect pests.

Some beneficial weeds repel insects and other pests through their smell , for example alliums and wormwood. Some weeds mask a companion plant's scent, or the pheromones of pest insects, as with ground ivy, as well as oregano and other mints.

Some also are unpleasant to small animals and ground insects, because of their spines or other features, keeping them away from an area to be protected.

Trap Crops

Some weeds act as trap crops, distracting pests away from valued plants. Insects often search for target plants by smell, and then land at random on anything green in the area of the scent. If they land on an edible "weed", they will stay there instead of going on to the intended victim. Sometimes, they actively prefer the trap crop.

Host-finding Disruption

Recent studies on host-plant finding have shown that flying pests are far less successful if their host-plants are surrounded by any other plant or even "decoy-plants" made of green plastic, cardboard, or any other green material.

- First, they seek plants by scent. Any "weed" that has a scent reduces the odds of them finding crop plants. Examples are Crow Garlic (wild chives) and Ground Ivy (a form of wild mint), both dramatically masking both plant scent and insect pheromones. They cut down Japanese beetle infestation, and caterpillar infestation, for example cabbage worm, tomato hornworm, and even squash bugs.

- Second, once an insect is near its target, it avoids landing on dirt, but lands on the nearest green thing. Bare earth gardening helps them home in perfectly on the victim crop. But if one is using "green mulch", even grass or clover, the odds are that they will make what's called an "inappropriate landing" on some green thing they don't want. They will then fly a short distance at random, and land on any other green thing. If they fail to accidentally hit the right kind of plant after several tries, they give up.

- If they plan to lay eggs on the crop, weeds provide one more line of defense: Even if they find the right plant, in order to ensure that they didn't hit on a dying plant or falling leaf, they then make short leaf-to-leaf flights before laying eggs. They must land on the "right kind of leaf" enough times in sequence, before they will risk laying their eggs. The more other greenery is nearby, the harder it is for them to remain on target and get enough reinforcement. Enough "inappropriate landings", and they give up, heading elsewhere.

One scientific study said that simply having clover growing nearby cut the odds of cabbage root flies hitting the right plant from 36% to 7%.

Companion Plants

Queen Anne's Lace provides shelter to nearby plants, as well as attracting predatory insects that eat pests like caterpillars, and may boost the productivity of tomato plants

Many plants can grow intercropped in the same space, because they exist on different levels in the same area, providing ground cover or working as a trellis for each other. This healthier style of horticulture is called forest gardening. Larger plants provide a wind break or shelter from noonday sun for more delicate plants.

Green Mulch

Conversely, some intercropped plants provide living mulch effect, used by inhibiting the growth of any weeds that are actually harmful, and creating a humid, cooler microclimate around nearby plants, stabilizing soil moisture more than they consume it for themselves.

Plants such as ryegrass, red clover, and white clover are examples of "weeds" that are living mulches, often welcomed in horticulture.

Herbicide

Repel plants or fungi, through a chemical means known as allelopathy. Specific other plants can be bothered by a chemical emission through their roots or air, slowing their growth, preventing seed germination, or even killing them.

Beneficial Insects

A common companion plant benefit from many weeds is to attract and provide habitat for beneficial insects or other organisms which benefit plants.

For example, wild umbellifers attract predatory wasps and flies. The adults eat nectar, but they feed common garden pests to their offspring .

Some weeds attract ladybugs or the "good" types of nematode, or provide ground cover for predatory beetles.

Uses for Humans

- Some beneficial weeds, such as lamb's quarters and purslane, are edible. This list of edible flowers includes many wildflowers that are considered weeds when not planted intentionally. Dandelion is an example of an edible weed.

- A number of weeds have been proposed as natural alternate sources for latex (rubber), including goldenrod, from which the tires were made on the car famously given by Henry Ford to Thomas Edison.

- Cocklebur and stinging nettle have been used for natural dyes.

- Milkweed is a more effective insulator than goose down.

- Some plants seem to subtly improve the flavor of other plants around them, for example stinging nettle, besides being edible if properly cooked, seems to increase essential oil production in nearby herbs.

Examples

- Clover is a legume. Like other beans, it hosts bacteria that fix nitrogen in the soil. Its vining nature covers the ground, sheltering more moisture than it consumes, providing a humid, cooler microclimate for surrounding plants as a "green mulch". It also is preferred by rodents over many garden crops, reducing the loss of vegetable crops.

- Dandelions possess a deep, strong tap root that breaks up hard soil, benefiting weaker-rooted plants nearby, and draw up nutrients from deeper than shallower-rooted nearby plants can access. They will also excrete minerals and nitrogen through their roots.

- Crow garlic, the wild chives found in sunny parts of a North American yard, has all of the companion plant benefits of other alliums, including repelling japanese beetles, aphids, and rodents, and being believed to benefit the flavor of solanums like tomatoes and peppers. It can be used as a substitute for garlic in cooking, though it may lend a bitter aftertaste.

- Bishop's lace (Queen Anne's Lace) works as a nurse plant for nearby crops like lettuce, shading them from overly intense sunlight and keeping more humidity in the air. It attracts predatory wasps and flies that eat vegetable pests. It has a scientifically tested beneficial effect on nearby tomato plants. When it is young it has an edible root, revealing its relationship to the domesticated carrot.

Agroforestry

Parkland in Burkina Faso: sorghum grown under Faidherbia albida and Borassus akeassii near Banfora

Agroforestry or agro-sylviculture is a land use management system in which trees or shrubs are grown around or among crops or pastureland. It combines shrubs and trees in agricultural and forestry technologies to create more diverse, productive, profitable, healthy, ecologically sound, and sustainable land-use systems.

As a Science

The theoretical base for agroforestry comes from ecology, via agroecology. From this perspective, agroforestry is one of the three principal land-use sciences. The other two are agriculture and forestry.

Agroforestry has a lot in common with intercropping. Both have two or more plant species (such as nitrogen-fixing plants) in close interaction, both provide multiple outputs, as a consequence, higher overall yields and, because a single application or input is shared, costs are reduced. Beyond these, there are gains specific to agroforestry.

Benefits

Agroforestry systems can be advantageous over conventional agricultural, and forest production methods. They can offer increased productivity, economic benefits, and more diversity in the ecological goods and services provided .(An example of this was seen in trying to conserve Milicia excelsa.)

Biodiversity in agroforestry systems is typically higher than in conventional agricultural systems. With two or more interacting plant species in a given land area, it creates a more complex habitat that can support a wider variety of birds, insects, and other animals. Depending upon the application, impacts of agroforestry can include:

- Reducing poverty through increased production of wood and other tree products for home consumption and sale
- Contributing to food security by restoring the soil fertility for food crops
- Cleaner water through reduced nutrient and soil runoff
- Countering global warming and the risk of hunger by increasing the number of drought-resistant trees and the subsequent production of fruits, nuts and edible oils
- Reducing deforestation and pressure on woodlands by providing farm-grown fuelwood
- Reducing or eliminating the need for toxic chemicals (insecticides, herbicides, etc.)
- Through more diverse farm outputs, improved human nutrition
- In situations where people have limited access to mainstream medicines, providing growing space for medicinal plants
- Increased crop stability
- Multifunctional site use i.e. crop production and animal grazing.
- Typically more drought resistant.
- Stabilises depleted soils from erosion
- Bioremediation

Agroforestry practices may also realize a number of other associated environmental goals, such as:

- Carbon sequestration
- Odour, dust, and noise reduction

- Green space and visual aesthetics

- Enhancement or maintenance of wildlife habitat

Adaptation to Climate Change

There is some evidence that, especially in recent years, poor smallholder farmers are turning to agroforestry as a mean to adapt to the impacts of climate change. A study from the CGIAR research program on Climate Change, Agriculture and Food Security (CCAFS) found from a survey of over 700 households in East Africa that at least 50% of those households had begun planting trees on their farms in a change from their practices 10 years ago. The trees ameliorate the effects of climate change by helping to stabilize erosion, improving water and soil quality and providing yields of fruit, tea, coffee, oil, fodder and medicinal products in addition to their usual harvest. Agroforestry was one of the most widely adopted adaptation strategies in the study, along with the use of improved crop varieties and intercropping.

Applications

Agroforestry represents a wide diversity in application and in practice. One listing includes over 50 distinct uses. The 50 or so applications can be roughly classified under a few broad headings. There are visual similarities between practices in different categories. This is expected as categorization is based around the problems addressed (countering winds, high rainfall, harmful insects, etc.) and the overall economic constraints and objectives (labor and other inputs costs, yield requirements, etc.). The categories include :

- Parklands

- Shade systems

- Crop-over-tree systems

- Alley cropping

- Strip cropping

- Fauna-based systems

- Boundary systems

- Taungyas

- Physical support systems

- Agroforests

- Wind break and shelterbelt.

Parkland

Parklands are visually defined by the presence of trees widely scattered over a large agricultural plot or pasture. The trees are usually of a single species with clear regional favorites. Among the beaks and benefits, the trees offer shade to grazing animals, protect crops against strong wind bursts, provide tree prunings for firewood, and are a roost for insect or rodent-eating birds.

There are other gains. Research with *Faidherbia albida* in Zambia showed that mature trees can sustain maize yields of 4.1 tonnes per hectare compared to 1.3 tonnes per hectare without these trees. Unlike other trees, Faidherbia sheds its nitrogen-rich leaves during the rainy crop growing season so it does not compete with the crop for light, nutrients and water. The leaves then regrow during the dry season and provide land cover and shade for crops.

Shade Systems

With shade applications, crops are purposely raised under tree canopies and within the resulting shady environment. For most uses, the understory crops are shade tolerant or the overstory trees have fairly open canopies. A conspicuous example is shade-grown coffee. This practice reduces weeding costs and improves the quality and taste of the coffee. Just because plants are grown under shade does not necessarily translate into lost or reduced yields. This is because the efficiency of photosynthesis drops off with increasing light intensity, and the rate of photosynthesis hardly increases once the light intensity is over about one tenth that of direct overhead sun. This means that plants under trees can still grow well even though they get less light. By having more than one level of vegetation, it is possible to get more photosynthesis, and overall yields, than with a single canopy layer.

Crop-over-tree Systems

Not commonly encountered, crop-over-tree systems employ woody perennials in the role of a cover crop. For this, small shrubs or trees pruned to near ground level are utilized. The purpose, as with any cover crop, is to increase in-soil nutrients and/or to reduce soil erosion.

Alley Cropping

Alley cropping corn fields between rows of walnut trees.

With alley cropping, crop strips alternate with rows of closely spaced tree or hedge species. Normally, the trees are pruned before planting the crop. The cut leafy material is spread over the crop area to provide nutrients for the crop. In addition to nutrients, the hedges serve as windbreaks and eliminate soil erosion.

Alley cropping has been shown to be advantageous in Africa, particularly in relation to improving maize yields in the sub-Saharan region. Use here relies upon the nitrogen fixing tree species *Sesbania sesban, euphorbia tricalii, Tephrosia vogelii, Gliricidia sepium* and *Faidherbia albida*. In one example, a ten-year experiment in Malawi showed that, by using the fertilizer tree Gliricidia (*Gliricidia sepium*) on land on which no mineral fertilizer was applied, maize yields averaged 3.3 tonnes per hectare as compared to one tonne per hectare in plots without fertilizer trees nor mineral fertilizers.

Strip Cropping

Strip cropping is similar to alley cropping in that trees alternate with crops. The difference is that, with alley cropping, the trees are in single row. With strip cropping, the trees or shrubs are planted in wide strip. The purpose can be, as with alley cropping, to provide nutrients, in leaf form, to the crop. With strip cropping, the trees can have a purely productive role, providing fruits, nuts, etc. while, at the same time, protecting nearby crops from soil erosion and harmful winds.

Fauna-based Systems

~ 1970   2004

Silvopasture over the years (Australia).

There are situations where trees benefit fauna. The most common examples are the silvopasture where cattle, goats, or sheep browse on grasses grown under trees. In hot climates, the animals are less stressed and put on weight faster when grazing in a cooler, shaded environment. Other variations have these animals directly eating the leaves of trees or shrubs.

There are similar systems for other types of fauna. Deer and hogs gain when living and feeding in a forest ecosystem, especially when the tree forage suits their dietary needs. Another variation, aqua-forestry, is where trees shade fish ponds. In many cases, the fish eat the leaves or fruit from the trees.

Boundary Systems

There are a number of applications that fall under the heading of a boundary system. These include the living fences, the riparian buffer, and windbreaks.

- A living fence can be a thick hedge or fencing wire strung on living trees. In addition to restricting the movement of people and animals, living fences offer habitat to insect-eating birds and, in the case of a boundary hedge, slow soil erosion.

- Riparian buffers are strips of permanent vegetation located along or near active watercourses or in ditches where water runoff concentrates. The purpose is to keep nutrients and soil from contaminating surface water.

- Windbreaks reduce the velocity of the winds over and around crops. This increases yields through reduced drying of the crop and/or by preventing the crop from toppling in strong wind gusts.

A riparian buffer bordering a river in Iowa.

Taungya

Taungya is a vastly used system originating in Burma. In the initial stages of an orchard or tree plantation, the trees are small and widely spaced. The free space between the newly planted trees can accommodate a seasonal crop. Instead of costly weeding, the underutilized area provides an additional output and income. More complex taungyas use the between-tree space for a series of crops. The crops become more shade resistant as the tree canopies grow and the amount of sunlight reaching the ground declines. If a plantation is thinned in the latter stages, this opens further the between-tree cropping opportunities.

Physical Support Systems

In the long history of agriculture, trellises are comparatively recent. Before this, grapes and other vine crops were raised atop pruned trees. Variations of the physical support theme depend upon the type of vine. The advantages come through greater in-field biodiversity. In many cases, the control of weeds, diseases, and insect pests are primary motives.

Agroforests

These are widely found in the humid tropics and are referenced by different names (forest gardening, forest farming, tropical home gardens and, where short-statured trees or shrubs dominate, shrub gardens). Through a complex, diverse mix of trees, shrubs, vines, and seasonal crops, these systems achieve the ecological dynamics of a forest ecosystem. Because of their internal ecology, they tend to be less susceptible to harmful insects, plant diseases, drought, and wind damage.

Historical Use

Agroforestry similar methods were historically utilized by Native Americans. California Indians would prescribe burn oak and other habitats to maintain a 'pyrodiversity collecting model'. This method allowed for greater health of trees and the habitat in general.

Challenges

Agroforestry is relevant to almost all environments and is a potential response to common problems around the globe, and agroforestry systems can be advantageous compared to conventional agriculture or forestry. Yet agroforestry is not very widespread, at least according to current but incomplete USDA surveys as of November, 2013.

As suggested by a survey of extension programs in the United States, some obstacles (ordered most critical to least critical) to agroforestry adoption include:

- Lack of developed markets for products
- Unfamiliarity with technologies
- Lack of awareness of successful agroforestry examples
- Competition between trees, crops, and animals
- Lack of financial assistance
- Lack of apparent profit potential
- Lack of demonstration sites
- Expense of additional management
- Lack of training or expertise
- Lack of knowledge about where to market products
- Lack of technical assistance
- Cannot afford adoption or start up costs, including costs of time
- Unfamiliarity with alternative marketing approaches (e.g. web)
- Unavailability of information about agroforestry
- Apparent inconvenience
- Lack of infrastructure (e.g. buildings, equipment)
- Lack of equipment
- Insufficient land
- Lack of seed/seedling sources
- Lack of scientific research

Some solutions to these obstacles have already been suggested although many depend on particular circumstances which vary from one location to the next.

Crop Rotation

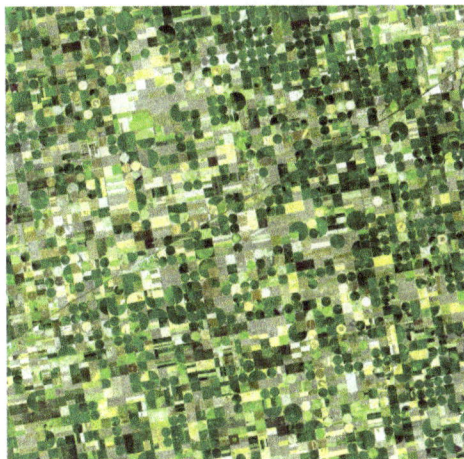

Satellite image of circular crop fields in Kansas in late June 2001. Healthy, growing crops are green. Corn would be growing into leafy stalks by then. Sorghum, which resembles corn, grows more slowly and would be much smaller and therefore, (possibly) paler. Wheat is a brilliant yellow as harvest occurs in June. Fields of brown have been recently harvested and plowed under or lie fallow for the year.

Effects of crop rotation and monoculture at the Swojec Experimental Farm, Wroclaw University of Environmental and Life Sciences. In the front field, the "Norfolk" crop rotation sequence (potatoes, oats, peas, rye) is being applied; in the back field, rye has been grown for 45 years in a row.

Crop rotation is the practice of growing a series of dissimilar or different types of crops in the same area in sequenced seasons. It is done so that the soil of farms is not used to only one type of nutrient. It helps in reducing soil erosion and increases soil fertility and crop yield.

Growing the same crop in the same place for many years in a row disproportionately depletes the soil of certain nutrients. With rotation, a crop that leaches the soil of one kind of nutrient is followed during the next growing season by a dissimilar crop that returns that nutrient to the soil or draws a different ratio of nutrients. In addition, crop rotation mitigates the buildup of pathogens and pests that often occurs when one species is continuously cropped, and can also improve soil structure and fertility by increasing biomass from varied root structures.

Crop rotation is used in both conventional and organic farming systems.

History

Agriculturalists have long recognized that suitable rotations – such as planting spring crops for livestock in place of grains for human consumption – make it possible to restore or to maintain a productive soil. Middle Eastern farmers practised crop rotation in 6000 BC without understanding the chemistry, alternately planting legumes and cereals. In the Bible, chapter 25 of the Book of Leviticus instructs the Israelites to observe a "Sabbath of the Land". Every seventh year they would not till, prune or even control insects. The Roman writer, Cato the Elder (234 – 149 BC), recommended that farmers "save carefully goat, sheep, cattle, and all other dung". From the times of Charlemagne (died 814), farmers in Europe transitioned from a two-field crop rotation to a three-field crop rotation. Under a two-field rotation, half the land was planted in a year, while the other half lay fallow. Then, in the next year, the two fields were reversed.

From the end of the Middle Ages until the 20th century, Europe's farmers practised three-field rotation, dividing available lands into three parts. One section was planted in the autumn with rye or winter wheat, followed by spring oats or barley; the second section grew crops such as peas, lentils, or beans; and the third field was left fallow. The three fields were rotated in this manner so that every three years, a field would rest and be fallow. Under the two-field system, if one has a total of 600 acres (2.4 km²) of fertile land, one would only plant 300 acres. Under the new three-field rotation system, one would plant (and therefore harvest) 400 acres. But the additional crops had a more significant effect than mere quantitative productivity. Since the spring crops were mostly legumes, they increased the overall nutrition of the people of Northern Europe.

Farmers in the region of Waasland (in present-day northern Belgium) pioneered a four-field rotation in the early 16th century, and the British agriculturist Charles Townshend (1674-1738) popularised this system in the 18th century. The sequence of four crops (wheat, turnips, barley and clover), included a fodder crop and a grazing crop, allowing livestock to be bred year-round. The four-field crop rotation became a key development in the British Agricultural Revolution. The rotation between arable and ley is sometimes called ley farming.

George Washington Carver (1860s - 1943) studied crop-rotation methods in the United States, teaching southern farmers to rotate soil-depleting crops like cotton with soil-enriching crops like peanuts and peas.

In the Green Revolution of the mid-20th century the traditional practice of crop rotation gave way in some parts of the world to the practice of supplementing the chemical inputs to the soil through topdressing with fertilizers, adding (for example) ammonium nitrate or urea and restoring soil pH with lime. Such practices aimed to increase yields, to prepare soil for specialist crops, and to reduce waste and inefficiency by simplifying planting and harvesting.

Crop Choice

A preliminary assessment of crop interrelationships can be found in how each crop: (1) contributes to soil organic matter (SOM) content, (2) provides for pest management, (3) manages deficient or excess nutrients, and (4) how it contributes to or controls for soil erosion.

Crop choice is often related to the goal the farmer is looking to achieve with the rotation, which could be weed management, increasing available nitrogen in the soil, controlling for erosion, or increasing soil structure and biomass, to name a few. When discussing crop rotations, crops are classified in different ways depending on what quality is being assessed: by family, by nutrient needs/benefits, and/or by profitability (i.e. cash crop versus cover crop). For example, giving adequate attention to plant family is essential to mitigating pests and pathogens. However, many farmers have success managing rotations by planning sequencing and cover crops around desirable cash crops. The following is a simplified classification based on crop quality and purpose.

Row Crops

Many crops which are critical for the market, like vegetables, are row crops (that is, grown in tight rows). While often the most profitable for farmers, these crops are more taxing on the soil Row crops typically have low biomass and shallow roots: this means the plant contributes low residue to the surrounding soil and has limited effects on structure. With much of the soil around the plant is exposed to disruption by rainfall and traffic, fields with row crops experience faster break down of organic matter by microbes, leaving fewer nutrients for future plants.

In short, while these crops may be profitable for the farm, they are nutrient depleting. Crop rotation practices exist to strike a balance between short-term profitability and long-term productivity.

Legumes

A great advantage of crop rotation comes from the interrelationship of nitrogen fixing-crops with nitrogen demanding crops. Legumes, like alfalfa and clover, collect available nitrogen from the soil in nodules on their root structure. When the plant is harvested, the biomass of uncollected roots breaks down, making the stored nitrogen available to future crops. Legumes are also a valued green manure: a crop that collects nutrients and fixes them at soil depths accessible to future crops.

In addition, legumes have heavy tap roots that burrow deep into the ground, lifting soil for better tilth and absorption of water.

Grasses and Cereals

Cereal and grasses are frequent cover crops because of the many advantages they supply to soil quality and structure. The dense and far-reaching root systems give ample structure to surrounding soil and provide significant biomass for soil organic matter.

Grasses and cereals are key in weed management as they compete with undesired plants for soil space and nutrients.

Green Manure

Green manure is a crop that is mixed into the soil. Both nitrogen-fixing legumes and nutrient scavengers, like grasses, can be used as green manure. Green manure of legumes is an excellent source of nitrogen, especially for organic systems, however, legume biomass doesn't contribute to lasting soil organic matter like grasses do.

Planning a Rotation

There are numerous factors that must be taken into consideration when planning a crop rotation. Planning an effective rotation requires weighing fixed and fluctuating production circumstances, including, but not limited to: market, farm size, labor supply, climate, soil type, growing practices, etc. Moreover, a crop rotation must consider in what condition one crop will leave the soil for the succeeding crop and how one crop can be seeded with another crop. For example, a nitrogen-fixing crop, like a legume, should always proceed a nitrogen depleting one; similarly, a low residue crop (i.e. a crop with low biomass) should be offset with a high biomass cover crop, like a mixture of grasses and legumes.

There is no limit to the number of crops that can be used in a rotation, or the amount of time a rotation takes to complete. Decisions about rotations are made years prior, seasons prior, or even at the very last minute when an opportunity to increase profits or soil quality presents itself. In short, there is no singular formula for rotation, but many considerations to take into account.

Implementation

Crop rotation systems may be enriched by the influences of other practices such as the addition of livestock and manure, intercropping or multiple cropping, and organic management low in pesticides and synthetic fertilizers.

Incorporation of Livestock

Introducing livestock makes the most efficient use of critical sod and cover crops; livestock (through manure) are able to distribute the nutrients in these crops throughout the soil rather than removing nutrients from the farm through the sale of hay. In systems where use of farm livestock would violate reservations growers or consumers may have about animal exploitation, efforts are made to surrogate this input through livestock in the soil, namely worms and microorganisms.

In Sub-Saharan Africa, as animal husbandry becomes less of a nomadic practice many herders have begun integrating crop production into their practice. This is known as mixed farming, or the practice of crop cultivation with the incorporation of raising cattle, sheep and/or goats by the same economic entity, is increasingly common. This interaction between the animal, the land and the crops are being done on a small scale all across this region. Crop residues provide animal feed, while the animals provide manure for replenishing crop nutrients and draft power. Both processes are extremely important in this region of the world as it is expensive and logistically unfeasible to transport in synthetic fertilizers and large-scale machinery. As an additional benefit, the cattle, sheep and/or goat provide milk and can act as a cash crop in the times of economic hardship.

Organic Farming

Crop rotation is a required practice in order for a farm to receive organic certification in the United States. The "Crop Rotation Practice Standard" for the National Organic Program under the U.S. Code of Federal Regulations, section §205.205, states that:

Farmers are required to implement a crop rotation that maintains or builds soil organic matter, works to control pests, manages and conserves nutrients, and protects against erosion. Producers of perennial crops that aren't rotated may utilize other practices, such as cover crops, to maintain soil health.

In addition to lowering the need for inputs by controlling for pests and weeds and increasing available nutrients, crop rotation helps organic growers increase the amount of biodiversity on their farms. Biodiversity is also a requirement of organic certification, however, there are no rules in place to regulate or reinforce this standard. Increasing the biodiversity of crops has beneficial effects on the surrounding ecosystem and can host a greater diversity of fauna, insects, and beneficial microorganism in the soil.< Some studies point to increased nutrient availability from crop rotation under organic systems compared to conventional practices as organic practices are less likely to inhibit of beneficial microbes in soil organic matter.

While multiple cropping and intercropping benefit from many of the same principals as crop rotation, they do not satisfy the requirement under the NOP.

Intercropping

Multiple cropping systems, such as intercropping or companion planting, offer more diversity and complexity within the same season or rotation, for example the three sisters. An example of companion planting is the inter-planting of corn with pole beans and vining squash or pumpkins. In this system, the beans provide nitrogen; the corn provides support for the beans and a "screen" against squash vine borer; the vining squash provides a weed suppressive canopy and discourages corn-hungry raccoons.

Double-cropping is common where two crops, typically of different species, are grown sequentially in the same growing season, or where one crop (e.g. vegetable) is grown continuously with a cover crop (e.g. wheat). This is advantageous for small farms, who often cannot afford to leave cover crops to replenish the soil for extended periods of time, as larger farms can. When multiple cropping is implemented on small farms, these systems can maximize benefits of crop rotation on available land resources.

Benefits

Agronomists describe the benefits to yield in rotated crops as "The Rotation Effect". There are many found benefits of rotation systems: however, there is no specific scientific basis for the sometimes 10-25% yield increase in a crop grown in rotation versus monoculture. The factors related to the increase are simply described as alleviation of the negative factors of monoculture cropping systems. Explanations due to improved nutrition; pest, pathogen, and weed stress reduction; and improved soil structure have been found in some cases to be correlated, but causation has not been determined for the majority of cropping systems.

Other benefits of rotation cropping systems include production cost advantages. Overall financial risks are more widely distributed over more diverse production of crops and/or livestock. Less reliance is placed on purchased inputs and over time crops can maintain production goals with fewer inputs. This in tandem with greater short and long term yields makes rotation a powerful tool for improving agricultural systems.

Soil Organic Matter

The use of different species in rotation allows for increased soil organic matter (SOM), greater soil structure, and improvement of the chemical and biological soil environment for crops. With more SOM, water infiltration and retention improves, providing increased drought tolerance and decreased erosion.

Soil organic matter is a mix of decaying material from biomass with active microorganisms. Crop rotation, by nature, increases exposure to biomass from sod, green manure, and a various other plant debris. The reduced need for intensive tillage under crop rotation allows biomass aggregation to lead to greater nutrient retention and utilization, decreasing the need for added nutrients. With tillage, disruption and oxidation of soil creates a less conducive environment for diversity and proliferation of microorganisms in the soil. These microorganisms are what make nutrients available to plants. So, where "active" soil organic matter is a key to productive soil, soil with low microbial activity provides significantly fewer nutrients to plants; this is true even though the quantity of biomass left in the soil may be the same.

Soil microorganisms also decrease pathogen and pest activity through competition. In addition, plants produce root exudates and other chemicals which manipulate their soil environment as well as their weed environment. Thus rotation allows increased yields from nutrient availability but also alleviation of allelopathy and competitive weed environments.

Carbon Sequestration

Studies have shown that crop rotations greatly increase soil organic carbon (SOC) content, the main constituent of soil organic matter. Carbon, along with hydrogen and oxygen, is a macro-nutrient for plants. Highly diverse rotations spanning long periods of time have shown to be even more effective in increasing SOC, while soil disturbances (e.g. from tillage) are responsible for exponential decline in SOC levels. In Brazil, conservation to no-till methods combined with intensive crop rotations has been shown an SOC sequestration rate of 0.41 tonnes per hectare per year.

In addition to enhancing crop productivity, sequestration of atmospheric carbon has great implications in reducing rates of climate change by removing carbon dioxide from the air.

Nitrogen Fixing

Rotating crops adds nutrients to the soil. Legumes, plants of the family Fabaceae, for instance, have nodules on their roots which contain nitrogen-fixing bacteria called rhizobia. It therefore makes good sense agriculturally to alternate them with cereals (family Poaceae) and other plants that require nitrates.

Pathogen and Pest Control

Crop rotation is also used to control pests and diseases that can become established in the soil over time. The changing of crops in a sequence decreases the population level of pests by (1) interrupting pest life cycles and (2) interrupting pest habitat. Plants within the same taxonomic family tend to have similar pests and pathogens. By regularly changing crops and keeping the soil occupied by cover crops instead of lying fallow, pest cycles can be broken or limited, especially cycles that benefit from overwintering in residue. For example, root-knot nematode is a serious problem for some plants in warm climates and sandy soils, where it slowly builds up to high levels in the soil, and can severely damage plant productivity by cutting off circulation from the plant roots. Growing a crop that is not a host for root-knot nematode for one season greatly reduces the level of the nematode in the soil, thus making it possible to grow a susceptible crop the following season without needing soil fumigation.

This principle is of particular use in organic farming, where pest control must be achieved without synthetic pesticides.

Weed Management

Integrating certain crops, especially cover crops, into crop rotations is of particular value to weed management. These crops crowd out weed through competition. In addition, the sod and compost from cover crops and green manure slows the growth of what weeds are still able to make it through the soil, giving the crops further competitive advantage. By removing slowing the growth and proliferation of weeds while cover crops are cultivated, farmers greatly reduce the presence of weeds for future crops, including shallow rooted and row crops, which are less resistant to weeds. Cover crops are, therefore, considered conservation crops because they protect otherwise fallow land from becoming overrun with weeds.

This system has advantages over other common practices for weeds management, such as tillage. Tillage is meant to inhibit growth of weeds by overturning the soil; however, this has a countering effect of exposing weed seeds that may have gotten buried and burying valuable crop seeds. Under crop rotation, the number of viable seeds in the soil is reduced through the reduction of the weed population.

Preventing Soil Erosion

Crop rotation can significantly reduce the amount of soil lost from erosion by water. In areas that are highly susceptible to erosion, farm management practices such as zero and reduced tillage can be supplemented with specific crop rotation methods to reduce raindrop impact, sediment detachment, sediment transport, surface runoff, and soil loss.

Protection against soil loss is maximized with rotation methods that leave the greatest mass of crop stubble (plant residue left after harvest) on top of the soil. Stubble cover in contact with the soil minimizes erosion from water by reducing overland flow velocity, stream power, and thus the ability of the water to detach and transport sediment. Soil Erosion and Cill prevent the disruption and detachment of soil aggregates that cause macropores to block, infiltration to decline, and runoff to increase. This significantly improves the resilience of soils when subjected to periods of erosion and stress.

The effect of crop rotation on erosion control varies by climate. In regions under relatively consistent climate conditions, where annual rainfall and temperature levels are assumed, rigid crop rotations can produce sufficient plant growth and soil cover. In regions where climate conditions are less predictable, and unexpected periods of rain and drought may occur, a more flexible approach for soil cover by crop rotation is necessary. An opportunity cropping system promotes adequate soil cover under these erratic climate conditions. In an opportunity cropping system, crops are grown when soil water is adequate and there is a reliable sowing window. This form of cropping system is likely to produce better soil cover than a rigid crop rotation because crops are only sown under optimal conditions, whereas rigid systems are not necessarily sown in the best conditions available.

Crop rotations also affect the timing and length of when a field is subject to fallow. This is very important because depending on a particular region's climate, a field could be the most vulnerable to erosion when it is under fallow. Efficient fallow management is an essential part of reducing erosion in a crop rotation system. Zero tillage is a fundamental management practice that promotes crop stubble retention under longer unplanned fallows when crops cannot be planted. Such management practices that succeed in retaining suitable soil cover in areas under fallow will ultimately reduce soil loss.

Biodiversity

Increasing the biodiversity of crops has beneficial effects on the surrounding ecosystem and can host a greater diversity of fauna, insects, and beneficial microorganisms in the soil. Some studies point to increased nutrient availability from crop rotation under organic systems compared to conventional practices as organic practices are less likely to inhibit of beneficial microbes in soil organic matter, such as arbuscular mycorrhizae, which increase nutrient uptake in plants. Increasing biodiversity also increases the resilience of agro-ecological systems.

Farm Productivity

Crop rotation contributes to increased yields through improved soil nutrition. By requiring planting and harvesting of different crops at different times, more land can be farmed with the same amount of machinery and labour.

Risk management

Different crops in the rotation can reduce the risks of adverse weather for the individual farmer.

Challenges

While crop rotation requires a great deal of planning, crop choice must respond to a number of fixed conditions (soil type, topography, climate, and irrigation) in addition to conditions that may change dramatically from year to the next (weather, market, labor supply). In this way, it is unwise to plan crops years in advance. Improper implementation of a crop rotation plan may lead to imbalances in the soil nutrient composition or a buildup of pathogens affecting a critical crop. The consequences of faulty rotation may take years to become apparent even to experienced soil scientists and can take just as long to correct.

Many challenges exist within the practices associated with crop rotation. For example, green manure from legumes can lead to an invasion of snails or slugs and the decay from green manure can occasionally suppress the growth of other crops.

Organic Farming

World map of organic agriculture (hectares)

Vegetables from ecological farming.

Organic farming is an alternative agricultural system which originated early in the 20th century in reaction to rapidly changing farming practices. Organic agriculture continues to be developed by various organic agriculture organizations today. It relies on fertilizers of organic origin such as compost, manure, green manure, and bone meal and places emphasis on techniques such as crop rotation and companion planting. Biological pest control, mixed cropping and the fostering of insect predators are encouraged. In general, organic standards are designed to allow the use of naturally occurring substances while prohibiting or strictly limiting synthetic substances. For instance, naturally occurring pesticides such as pyrethrin and rotenone are permitted, while synthetic fertilizers and pesticides are generally prohibited. Synthetic substances that are allowed include, for example, copper sulfate, elemental sulfur and Ivermectin. Genetically modified organisms, nanomaterials, human sewage sludge, plant growth regulators, hormones, and antibiotic use in livestock husbandry are prohibited. Reasons for advocation of organic farming include real or perceived advantages in sustainability, openness, self-sufficiency, autonomy/independence, health, food security, and food safety, although the match between perception and reality is continually challenged.

Organic agricultural methods are internationally regulated and legally enforced by many nations, based in large part on the standards set by the International Federation of Organic Agriculture Movements (IFOAM), an international umbrella organization for organic farming organizations established in 1972. Organic agriculture can be defined as:

an integrated farming system that strives for sustainability, the enhancement of soil fertility and biological diversity whilst, with rare exceptions, prohibiting synthetic pesticides, antibiotics, synthetic fertilizers, genetically modified organisms, and growth hormones.

Since 1990 the market for organic food and other products has grown rapidly, reaching $63 billion worldwide in 2012. This demand has driven a similar increase in organically managed farmland that grew from 2001 to 2011 at a compounding rate of 8.9% per annum. As of 2011, approximately 37,000,000 hectares (91,000,000 acres) worldwide were farmed organically, representing approximately 0.9 percent of total world farmland.

History

Agriculture was practiced for thousands of years without the use of artificial chemicals. Artificial fertilizers were first created during the mid-19th century. These early fertilizers were cheap, powerful, and easy to transport in bulk. Similar advances occurred in chemical pesticides in the 1940s, leading to the decade being referred to as the 'pesticide era'. These new agricultural techniques, while beneficial in the short term, had serious longer term side effects such as soil compaction, erosion, and declines in overall soil fertility, along with health concerns about toxic chemicals entering the food supply. In the late 1800s and early 1900s, soil biology scientists began to seek ways to remedy these side effects while still maintaining higher production.

Biodynamic agriculture was the first modern system of agriculture to focus exclusively on organic methods. Its development began in 1924 with a series of eight lectures on agriculture given by Rudolf Steiner. These lectures, the first known presentation of what later came to be known as organic agriculture, were held in response to a request by farmers who noticed degraded soil conditions and a deterioration in the health and quality of crops and livestock resulting from the use of chemical fertilizers. The one hundred eleven attendees, less than half of whom were farmers, came from six countries, primarily Germany and Poland. The lectures were published in November 1924; the first English translation appeared in 1928 as *The Agriculture Course*.

In 1921, Albert Howard and his wife Gabrielle Howard, accomplished botanists, founded an Institute of Plant Industry to improve traditional farming methods in India. Among other things, they brought improved implements and improved animal husbandry methods from their scientific training; then by incorporating aspects of the local traditional methods, developed protocalls for the rotation of crops, erosion prevention techniques, and the systematic use of composts and manures. Stimulated by these experiences of traditional farming, when Albert Howard returned to Britain in the early 1930s he began to promulgate a system of natural agriculture.

In July 1939, Ehrenfried Pfeiffer, the author of the standard work on biodynamic agriculture (*Bio-Dynamic Farming and Gardening*), came to the UK at the invitation of Walter James, 4th Baron Northbourne as a presenter at the Betteshanger Summer School and Conference on Biodynamic Farming at Northbourne's farm in Kent. One of the chief purposes of the conference was to bring together the proponents of various approaches to organic agriculture in order that they might cooperate within a larger movement. Howard attended the conference, where he met Pfeiffer. In the following year, Northbourne published his manifesto of organic farming, *Look to the Land*, in which he coined the term "organic farming." The Betteshanger conference has been described as the 'missing link' between biodynamic agriculture and other forms of organic farming.

In 1940 Howard published his *An Agricultural Testament*. In this book he adopted Northbourne's terminology of "organic farming." Howard's work spread widely, and he became known as the "father of organic farming" for his work in applying scientific knowledge and principles to various traditional and natural methods. In the United States J.I. Rodale, who was keenly interested both in Howard's ideas and in biodynamics, founded in the 1940s both a working organic farm for trials and experimentation, The Rodale Institute, and the Rodale Press to teach and advocate organic methods to the wider public. These became important influences on the spread of organic agriculture. Further work was done by Lady Eve Balfour in the United Kingdom, and many others across the world.

Increasing environmental awareness in the general population in modern times has transformed the originally supply-driven organic movement to a demand-driven one. Premium prices and some government subsidies attracted farmers. In the developing world, many producers farm according to traditional methods that are comparable to organic farming, but not certified, and that may not include the latest scientific advancements in organic agriculture. In other cases, farmers in the developing world have converted to modern organic methods for economic reasons.

Terminology

Biodynamic agriculturists, who based their work on Steiner's spiritually-oriented anthroposophy, used the term "organic" to indicate that a farm should be viewed as a living organism, in the sense of the following quotation:

"An organic farm, properly speaking, is not one that uses certain methods and substances and avoids others; it is a farm whose structure is formed in imitation of the structure of a natural system that has the integrity, the independence and the benign dependence of an organism"

— *Wendell Berry, "The Gift of Good Land"*

The use of "organic" popularized by Howard and Rodale, on the other hand, refers more narrowly to the use of organic matter derived from plant compost and animal manures to improve the humus content of soils, grounded in the work of early soil scientists who developed what was then called "humus farming." Since the early 1940s the two camps have tended to merge.

Methods

Organic cultivation of mixed vegetables in Capay, California. Note the hedgerow in the background.

"Organic agriculture is a production system that sustains the health of soils, ecosystems and people. It relies on ecological processes, biodiversity and cycles adapted to local conditions, rather than the use of inputs with adverse effects. Organic agriculture combines tradition, innovation and science to benefit the shared environment and promote fair relationships and a good quality of life for all involved..."

— International Federation of Organic Agriculture Movements

Organic farming methods combine scientific knowledge of ecology and modern technology with traditional farming practices based on naturally occurring biological processes. Organic farming methods are studied in the field of agroecology. While conventional agriculture uses synthetic pesticides and water-soluble synthetically purified fertilizers, organic farmers are restricted by regulations to using natural pesticides and fertilizers. An example of a natural pesticide is pyrethrin, which is found naturally in the Chrysanthemum flower. The principal methods of organic farming include crop rotation, green manures and compost, biological pest control, and mechanical cultivation. These measures use the natural environment to enhance agricultural productivity: legumes are planted to fix nitrogen into the soil, natural insect predators are encouraged, crops are rotated to confuse pests and renew soil, and natural materials such as potassium bicarbonate and mulches are used to control disease and weeds. Genetically modified seeds and animals are excluded.

While organic is fundamentally different from conventional because of the use of carbon based fertilizers compared with highly soluble synthetic based fertilizers and biological pest control instead of synthetic pesticides, organic farming and large-scale conventional farming are not entirely mutually exclusive. Many of the methods developed for organic agriculture have been borrowed by more conventional agriculture. For example, Integrated Pest Management is a multifaceted strategy that uses various organic methods of pest control whenever possible, but in conventional farming could include synthetic pesticides only as a last resort.

Crop Diversity

Organic farming encourages Crop diversity. The science of agroecology has revealed the benefits of polyculture (multiple crops in the same space), which is often employed in organic farming. Planting a variety of vegetable crops supports a wider range of beneficial insects, soil microorganisms, and other factors that add up to overall farm health. Crop diversity helps environments thrive and protects species from going extinct.

Soil Management

Organic farming relies heavily on the natural breakdown of organic matter, using techniques like green manure and composting, to replace nutrients taken from the soil by previous crops. This biological process, driven by microorganisms such as mycorrhiza, allows the natural production of nutrients in the soil throughout the growing season, and has been referred to as *feeding the soil to feed the plant*. Organic farming uses a variety of methods to improve soil fertility, including crop rotation, cover cropping, reduced tillage, and application of compost. By reducing tillage, soil is not inverted and exposed to air; less carbon is lost to the atmosphere resulting in more soil organic carbon. This has an added benefit of carbon sequestration, which can reduce green house gases and help reverse climate change.

Plants need nitrogen, phosphorus, and potassium, as well as micronutrients and symbiotic relationships with fungi and other organisms to flourish, but getting enough nitrogen, and particularly synchronization so that plants get enough nitrogen at the right time (when plants need it most), is a challenge for organic farmers. Crop rotation and green manure ("cover crops") help to provide nitrogen through legumes (more precisely, the *Fabaceae* family), which fix nitrogen from the atmosphere through symbiosis with rhizobial bacteria. Intercropping, which is sometimes used for insect and disease control, can also increase soil nutrients, but the competition between the legume and the crop can be problematic and wider spacing between crop rows is required. Crop residues can be ploughed back into the soil, and different plants leave different amounts of nitrogen, potentially aiding synchronization. Organic farmers also use animal manure, certain processed fertilizers such as seed meal and various mineral powders such as rock phosphate and green sand, a naturally occurring form of potash that provides potassium. Together these methods help to control erosion. In some cases pH may need to be amended. Natural pH amendments include lime and sulfur, but in the U.S. some compounds such as iron sulfate, aluminum sulfate, magnesium sulfate, and soluble boron products are allowed in organic farming.

Mixed farms with both livestock and crops can operate as ley farms, whereby the land gathers fertility through growing nitrogen-fixing forage grasses such as white clover or alfalfa and grows cash crops or cereals when fertility is established. Farms without livestock ("stockless") may find it more difficult to maintain soil fertility, and may rely more on external inputs such as imported manure as well as grain legumes and green manures, although grain legumes may fix limited nitrogen because they are harvested. Horticultural farms that grow fruits and vegetables in protected conditions often relay even more on external inputs.

Biological research into soil and soil organisms has proven beneficial to organic farming. Varieties of bacteria and fungi break down chemicals, plant matter and animal waste into productive soil nutrients. In turn, they produce benefits of healthier yields and more productive soil for future crops. Fields with less or no manure display significantly lower yields, due to decreased soil microbe community. Increased manure improves biological activity, providing a healthier, more arable soil system and higher yields.

Weed Management

Organic weed management promotes weed suppression, rather than weed elimination, by enhancing crop competition and phytotoxic effects on weeds. Organic farmers integrate cultural, biological, mechanical, physical and chemical tactics to manage weeds without synthetic herbicides.

Organic standards require rotation of annual crops, meaning that a single crop cannot be grown in the same location without a different, intervening crop. Organic crop rotations frequently include weed-suppressive cover crops and crops with dissimilar life cycles to discourage weeds associated with a particular crop. Research is ongoing to develop organic methods to promote the growth of natural microorganisms that suppress the growth or germination of common weeds.

Other cultural practices used to enhance crop competitiveness and reduce weed pressure include selection of competitive crop varieties, high-density planting, tight row spacing, and late planting into warm soil to encourage rapid crop germination.

Mechanical and physical weed control practices used on organic farms can be broadly grouped as:

- Tillage - Turning the soil between crops to incorporate crop residues and soil amendments; remove existing weed growth and prepare a seedbed for planting; turning soil after seeding to kill weeds, including cultivation of row crops;

- Mowing and cutting - Removing top growth of weeds;

- Flame weeding and thermal weeding - Using heat to kill weeds; and

- Mulching - Blocking weed emergence with organic materials, plastic films, or landscape fabric.

Some critics, citing work published in 1997 by David Pimentel of Cornell University, which described an epidemic of soil erosion worldwide, have raised concerned that tillage contribute to the erosion epidemic. The FAO and other organizations have advocated a 'no-till' approach to both conventional and organic farming, and point out in particular that crop rotation techniques used in organic farming are excellent no-till approaches. A study published in 2005 by Pimentel and colleagues confirmed that 'Crop rotations and cover cropping (green manure) typical of organic agriculture reduce soil erosion, pest problems, and pesticide use.' Some naturally sourced chemicals are allowed for herbicidal use. These include certain formulations of acetic acid (concentrated vinegar), corn gluten meal, and essential oils. A few selective bioherbicides based on fungal pathogens have also been developed. At this time, however, organic herbicides and bioherbicides play a minor role in the organic weed control toolbox.

Weeds can be controlled by grazing. For example, geese have been used successfully to weed a range of organic crops including cotton, strawberries, tobacco, and corn, reviving the practice of keeping cotton patch geese, common in the southern U.S. before the 1950s. Similarly, some rice farmers introduce ducks and fish to wet paddy fields to eat both weeds and insects.

Controlling Other Organisms

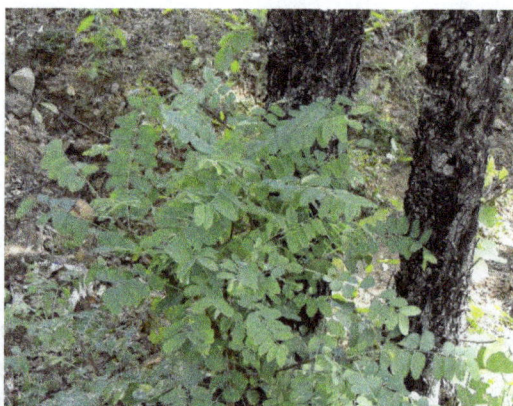

Chloroxylon is used for Pest Management in Organic Rice Cultivation in Chhattisgarh, India

Organisms aside from weeds that cause problems on organic farms include arthropods (e.g., insects, mites), nematodes, fungi and bacteria. Organic practices include, but are not limited to:

- encouraging predatory beneficial insects to control pests by serving them nursery plants and/or an alternative habitat, usually in a form of a shelterbelt, hedgerow, or beetle bank;

- encouraging beneficial microorganisms;

- rotating crops to different locations from year to year to interrupt pest reproduction cycles;

- planting companion crops and pest-repelling plants that discourage or divert pests;

- using row covers to protect crops during pest migration periods;

- using biologic pesticides and herbicides

- using stale seed beds to germinate and destroy weeds before planting

- using sanitation to remove pest habitat;

- Using insect traps to monitor and control insect populations.

- Using physical barriers, such as row covers

Examples of predatory beneficial insects include minute pirate bugs, big-eyed bugs, and to a lesser extent ladybugs (which tend to fly away), all of which eat a wide range of pests. Lacewings are also effective, but tend to fly away. Praying mantis tend to move more slowly and eat less heavily. Parasitoid wasps tend to be effective for their selected prey, but like all small insects can be less effective outdoors because the wind controls their movement. Predatory mites are effective for controlling other mites.

Naturally derived insecticides allowed for use on organic farms use include *Bacillus thuringiensis* (a bacterial toxin), pyrethrum (a chrysanthemum extract), spinosad (a bacterial metabolite), neem (a tree extract) and rotenone (a legume root extract). Fewer than 10% of organic farmers use these pesticides regularly; one survey found that only 5.3% of vegetable growers in California use rotenone while 1.7% use pyrethrum. These pesticides are not always more safe or environmentally friendly than synthetic pesticides and can cause harm. The main criterion for organic pesticides is that they are naturally derived, and some naturally derived substances have been controversial. Controversial natural pesticides include rotenone, copper, nicotine sulfate, and pyrethrums Rotenone and pyrethrum are particularly controversial because they work by attacking the nervous system, like most conventional insecticides. Rotenone is extremely toxic to fish and can induce symptoms resembling Parkinson's disease in mammals. Although pyrethrum (natural pyrethrins) is more effective against insects when used with piperonyl butoxide (which retards degradation of the pyrethrins), organic standards generally do not permit use of the latter substance.

Naturally derived fungicides allowed for use on organic farms include the bacteria *Bacillus subtilis* and *Bacillus pumilus*; and the fungus *Trichoderma harzianum*. These are mainly effective for diseases affecting roots. Compost tea contains a mix of beneficial microbes, which may attack or out-compete certain plant pathogens, but variability among formulations and preparation methods may contribute to inconsistent results or even dangerous growth of toxic microbes in compost teas.

Some naturally derived pesticides are not allowed for use on organic farms. These include nicotine sulfate, arsenic, and strychnine.

Synthetic pesticides allowed for use on organic farms include insecticidal soaps and horticultural oils for insect management; and Bordeaux mixture, copper hydroxide and sodium bicarbonate for managing fungi. Copper sulfate and Bordeaux mixture (copper sulfate plus lime), approved for organic use in various jurisdictions, can be more environmentally problematic than some synthetic fungicides dissallowed in organic farming Similar concerns apply to copper hydroxide. Repeated application of copper sulfate or copper hydroxide as a fungicide may eventually result in copper accumulation to toxic levels in soil, and admonitions to avoid excessive accumulations of copper in soil appear in various organic standards and elsewhere. Environmental concerns for several kinds of biota arise at average rates of use of such substances for some crops. In the European Union, where replacement of copper-based fungicides in organic agriculture is a policy priority, research is seeking alternatives for organic production.

Livestock

For livestock like these healthy cows vaccines play an important part in animal health since antibiotic therapy is prohibited in organic farming

Raising livestock and poultry, for meat, dairy and eggs, is another traditional farming activity that complements growing. Organic farms attempt to provide animals with natural living conditions and feed. Organic certification verifies that livestock are raised according to the USDA organic regulations throughout their lives. These regulations include the requirement that all animal feed must be certified organic.

Organic livestock may be, and must be, treated with medicine when they are sick, but drugs cannot be used to promote growth, their feed must be organic, and they must be pastured.

Also, horses and cattle were once a basic farm feature that provided labor, for hauling and plowing, fertility, through recycling of manure, and fuel, in the form of food for farmers and other animals. While today, small growing operations often do not include livestock, domesticated animals are a desirable part of the organic farming equation, especially for true sustainability, the ability of a farm to function as a self-renewing unit.

Genetic Modification

A key characteristic of organic farming is the rejection of genetically engineered plants and animals. On 19 October 1998, participants at IFOAM's 12th Scientific Conference issued the Mar del Plata Declaration, where more than 600 delegates from over 60 countries voted unanimously to exclude the use of genetically modified organisms in food production and agriculture.

Although opposition to the use of any transgenic technologies in organic farming is strong, agricultural researchers Luis Herrera-Estrella and Ariel Alvarez-Morales continue to advocate integration of transgenic technologies into organic farming as the optimal means to sustainable agriculture, particularly in the developing world, as does author and scientist Pamela Ronald, who views this kind of biotechnology as being consistent with organic principles.

Although GMOs are excluded from organic farming, there is concern that the pollen from genetically modified crops is increasingly penetrating organic and heirloom seed stocks, making it difficult, if not impossible, to keep these genomes from entering the organic food supply. Differing regulations among countries limits the availability of GMOs to certain countries, as described in the article on regulation of the release of genetic modified organisms.

Tools

Organic farmers use a number of traditional farm tools to do farming. Due to the goals of sustainability in organic farming, organic farmers try to minimize their reliance on fossil fuels. In the developing world on small organic farms tools are normally constrained to hand tools and diesel powered water pumps. A recent study evaluated the use of open-source 3-D printers (called RepRaps using a bioplastic polylactic acid (PLA) on organic farms. PLA is a strong biodegradable and recyclable thermoplastic appropriate for a range of representative products in five categories of prints: handtools, food processing, animal management, water management and hydroponics. Such open source hardware is attractive to all types of small farmers as it provides control for farmers over their own equipment; this is exemplified by Open Source Ecology, Farm Hack and FarmBot.

Standards

Standards regulate production methods and in some cases final output for organic agriculture. Standards may be voluntary or legislated. As early as the 1970s private associations certified organic producers. In the 1980s, governments began to produce organic production guidelines. In the 1990s, a trend toward legislated standards began, most notably with the 1991 EU-Eco-regulation developed for European Union, which set standards for 12 countries, and a 1993 UK program. The EU's program was followed by a Japanese program in 2001, and in 2002 the U.S. created the National Organic Program (NOP). As of 2007 over 60 countries regulate organic farming (IFOAM 2007:11). In 2005 IFOAM created the Principles of Organic Agriculture, an international guideline for certification criteria. Typically the agencies accredit certification groups rather than individual farms.

Organic production materials used in and foods are tested independently by the Organic Materials Review Institute.

Composting

Using manure as a fertiliser risks contaminating food with animal gut bacteria, including pathogenic strains of E. coli that have caused fatal poisoning from eating organic food. To combat this risk, USDA organic standards require that manure must be sterilized through high temperature thermophilic composting. If raw animal manure is used, 120 days must pass before the crop is harvested if the final product comes into direct contact with the soil. For products that don't directly contact soil, 90 days must pass prior to harvest.

Economics

The economics of organic farming, a subfield of agricultural economics, encompasses the entire process and effects of organic farming in terms of human society, including social costs, opportunity costs, unintended consequences, information asymmetries, and economies of scale. Although the scope of economics is broad, agricultural economics tends to focus on maximizing yields and efficiency at the farm level. Economics takes an anthropocentric approach to the value of the natural world: biodiversity, for example, is considered beneficial only to the extent that it is valued by people and increases profits. Some entities such as the European Union subsidize organic farming, in large part because these countries want to account for the externalities of reduced water use, reduced water contamination, reduced soil erosion, reduced carbon emissions, increased biodiversity, and assorted other benefits that result from organic farming.

Traditional organic farming is labor and knowledge-intensive whereas conventional farming is capital-intensive, requiring more energy and manufactured inputs.

Organic farmers in California have cited marketing as their greatest obstacle.

Geographic Producer Distribution

The markets for organic products are strongest in North America and Europe, which as of 2001 are estimated to have $6 and $8 billion respectively of the $20 billion global market. As of 2007 Australasia has 39% of the total organic farmland, including Australia's 1,180,000 hectares (2,900,000 acres) but 97 percent of this land is sprawling rangeland (2007:35). US sales are 20x as much. Europe farms 23 percent of global organic farmland (6,900,000 ha (17,000,000 acres)), followed by Latin America with 19 percent (5.8 million hectares - 14.3 million acres). Asia has 9.5 percent while North America has 7.2 percent. Africa has 3 percent.

Besides Australia, the countries with the most organic farmland are Argentina (3.1 million hectares - 7.7 million acres), China (2.3 million hectares - 5.7 million acres), and the United States (1.6 million hectares - 4 million acres). Much of Argentina's organic farmland is pasture, like that of Australia (2007:42). Spain, Germany, Brazil (the world's largest agricultural exporter), Uruguay, and the UK follow the United States in the amount of organic land (2007:26).

In the European Union (EU25) 3.9% of the total utilized agricultural area was used for organic production in 2005. The countries with the highest proportion of organic land were Austria (11%) and Italy (8.4%), followed by the Czech Republic and Greece (both 7.2%). The lowest figures were shown for Malta (0.1%), Poland (0.6%) and Ireland (0.8%). In 2009, the proportion of organic land in the EU grew to 4.7%. The countries with highest share of agricultural land were Liechtenstein (26.9%), Austria (18.5%) and Sweden (12.6%). 16% of all farmers in Austria produced organically in 2010. By the same year the proportion of organic land increased to 20%.: In 2005 168,000 ha (415,000 ac) of land in Poland was under organic management. In 2012, 288,261 hectares (712,308 acres) were under organic production, and there were about 15,500 organic farmers;

retail sales of organic products were EUR 80 million in 2011. As of 2012 organic exports were part of the government's economic development strategy.

After the collapse of the Soviet Union in 1991, agricultural inputs that had previously been purchased from Eastern bloc countries were no longer available in Cuba, and many Cuban farms converted to organic methods out of necessity. Consequently, organic agriculture is a mainstream practice in Cuba, while it remains an alternative practice in most other countries. Cuba's organic strategy includes development of genetically modified crops; specifically corn that is resistant to the palomilla moth

Growth

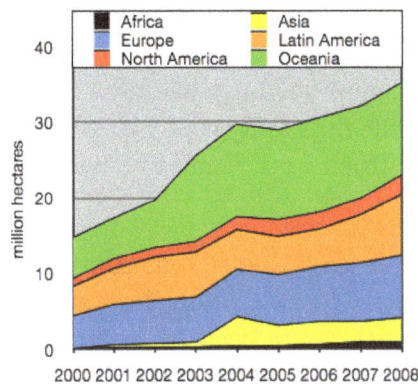

Organic farmland by world region (2000-2008)

In 2001, the global market value of certified organic products was estimated at USD $20 billion. By 2002, this was USD $23 billion and by 2015 more than USD $43 billion. By 2014, retail sales of organic products reached USD $80 billion worldwide. North America and Europe accounted for more than 90% of all organic product sales.

Organic agricultural land increased almost fourfold in 15 years, from 11 million hectares in 1999 to 43.7 million hectares in 2014. Between 2013 and 2014, organic agricultural land grew by 500,000 hectares worldwide, increasing in every region except Latin America. During this time period, Europe's organic farmland increased 260,000 hectares to 11.6 million total (+2.3%), Asia's increased 159,000 hectares to 3.6 million total (+4.7%), Africa's increased 54,000 hectares to 1.3 million total (+4.5%), and North America's increased 35,000 hectares to 3.1 million total (+1.1%). As of 2014, the country with the most organic land was Australia (17.2 million hectares), followed by Argentina (3.1 million hectares), and the United States (2.2 million hectares).

In 2013, the number of organic producers grew by almost 270,000, or more than 13%. By 2014, there were a reported 2.3 million organic producers in the world. Most of the total global increase took place in the Philippines, Peru, China, and Thailand. Overall, the majority of all organic producers are in India (650,000 in 2013), Uganda (190,552 in 2014), Mexico (169,703 in 2013) and the Philippines (165,974 in 2014).

Productivity

Studies comparing yields have had mixed results. These differences among findings can often be attributed to variations between study designs including differences in the crops studied and the methodology by which results were gathered.

A 2012 meta-analysis found that productivity is typically lower for organic farming than conventional farming, but that the size of the difference depends on context and in some cases may be very small. While organic yields can be lower than conventional yields, another meta-analysis published in Sustainable Agriculture Research in 2015, concluded that certain organic on-farm practices could help narrow this gap. Timely weed management and the application of manure in conjunction with legume forages/cover crops were shown to have positive results in increasing organic corn and soybean productivity. More experienced organic farmers were also found to have higher yields than other organic farmers who were just starting out.

Another meta-analysis published in the journal Agricultural Systems in 2011 analyzed 362 datasets and found that organic yields were on average 80% of conventional yields. The author's found that there are relative differences in this yield gap based on crop type with crops like soybeans and rice scoring higher than the 80% average and crops like wheat and potato scoring lower. Across global regions, Asia and Central Europe were found to have relatively higher yields and Northern Europe relatively lower than the average.

A 2007 study compiling research from 293 different comparisons into a single study to assess the overall efficiency of the two agricultural systems has concluded that "organic methods could produce enough food on a global per capita basis to sustain the current human population, and potentially an even larger population, without increasing the agricultural land base." The researchers also found that while in developed countries, organic systems on average produce 92% of the yield produced by conventional agriculture, organic systems produce 80% more than conventional farms in developing countries, because the materials needed for organic farming are more accessible than synthetic farming materials to farmers in some poor countries. This study was strongly contested by another study published in 2008, which stated, and was entitled, "Organic agriculture cannot feed the world" and said that the 2007 came up with "a major overestimation of the productivity of OA" "because data are misinterpreted and calculations accordingly are erroneous." Additional research needs to be conducted in the future to further clarify these claims.

Long Term Studies

A study published in 2005 compared conventional cropping, organic animal-based cropping, and organic legume-based cropping on a test farm at the Rodale Institute over 22 years. The study found that "the crop yields for corn and soybeans were similar in the organic animal, organic legume, and conventional farming systems". It also found that "significantly less fossil energy was expended to produce corn in the Rodale Institute's organic animal and organic legume systems than in the conventional production system. There was little difference in energy input between the different treatments for producing soybeans. In the organic systems, synthetic fertilizers and pesticides were generally not used". As of 2013 the Rodale study was ongoing and a thirty-year anniversary report was published by Rodale in 2012.

A long-term field study comparing organic/conventional agriculture carried out over 21 years in Switzerland concluded that "Crop yields of the organic systems averaged over 21 experimental years at 80% of the conventional ones. The fertilizer input, however, was 34 − 51% lower, indicating an efficient production. The organic farming systems used 20 – 56% less energy to produce a crop unit and per land area this difference was 36 – 53%. In spite of the considerably lower pesticide input the quality of organic products was hardly discernible from conventional analytically and even came off better in food preference trials and picture creating methods"

Profitability

In the United States, organic farming has been shown to be 2.9 to 3.8 times more profitable for the farmer than conventional farming when prevailing price premiums are taken into account. Globally, organic farming is between 22 and 35 percent more profitable for farmers than conventional methods, according to a 2015 meta-analysis of studies conducted across five continents.

The profitability of organic agriculture can be attributed to a number of factors. First, organic farmers do not rely on synthetic fertilizer and pesticide inputs, which can be costly. In addition, organic foods currently enjoy a price premium over conventionally produced foods, meaning that organic farmers can often get more for their yield.

The price premium for organic food is an important factor in the economic viability of organic farming. In 2013 there was a 100% price premium on organic vegetables and a 57% price premium for organic fruits. These percentages are based on wholesale fruit and vegetable prices, available through the United States Department of Agriculture's Economic Research Service. Price premiums exist not only for organic versus nonorganic crops, but may also vary depending on the venue where the product is sold: farmers markets, grocery stores, or wholesale to restaurants. For many producers, direct sales at farmers markets are most profitable because the farmer receives the entire markup, however this is also the most time and labor-intensive approach.

There have been signs of organic price premiums narrowing in recent years, which lowers the economic incentive for farmers to convert to or maintain organic production methods. Data from 22 years of experiments at the Rodale Institute found that, based on the current yields and production costs associated with organic farming in the United States, a price premium of only 10% is required to achieve parity with conventional farming. A separate study found that on a global scale, price premiums of only 5-7% percent were needed to break even with conventional methods. Without the price premium, profitability for farmers is mixed.

For markets and supermarkets organic food is profitable as well, and is generally sold at significantly higher prices than non-organic food.

Energy Efficiency

In the most recent assessments of the energy efficiency of organic versus conventional agriculture, results have been mixed regarding which form is more carbon efficient. Organic farm systems have more often than not been found to be more energy efficient, however, this is not always the case. More than anything, results tend to depend upon crop type and farm size.

A comprehensive comparison of energy efficiency in grain production, produce yield, and animal husbandry concluded that organic farming had a higher yield per unit of energy over the vast majority of the crops and livestock systems. For example, two studies - both comparing organically-versus conventionally-farmed apples - declare contradicting results, one saying organic farming is more energy efficient, the other saying conventionally is more efficient.

It has generally been found that the labor input per unit of yield was higher for organic systems compared with conventional production.

Sales and Marketing

Most sales are concentrated in developed nations. In 2008, 69% of Americans claimed to occasionally buy organic products, down from 73% in 2005. One theory for this change was that consumers were substituting "local" produce for "organic" produce.

Distributors

The USDA requires that distributors, manufacturers, and processors of organic products be certified by an accredited state or private agency. In 2007, there were 3,225 certified organic handlers, up from 2,790 in 2004.

Organic handlers are often small firms; 48% reported sales below $1 million annually, and 22% between $1 and $5 million per year. Smaller handlers are more likely to sell to independent natural grocery stores and natural product chains whereas large distributors more often market to natural product chains and conventional supermarkets, with a small group marketing to independent natural product stores. Some handlers work with conventional farmers to convert their land to organic with the knowledge that the farmer will have a secure sales outlet. This lowers the risk for the handler as well as the farmer. In 2004, 31% of handlers provided technical support on organic standards or production to their suppliers and 34% encouraged their suppliers to transition to organic. Smaller farms often join together in cooperatives to market their goods more effectively.

93% of organic sales are through conventional and natural food supermarkets and chains, while the remaining 7% of U.S. organic food sales occur through farmers' markets, foodservices, and other marketing channels.

Direct-to-Consumer Sales

In the 2012 Census, direct-to-consumer sales equaled $1.3 billion, up from $812 million in 2002, an increase of 60 percent. The number of farms that utilize direct-to-consumer sales was 144,530 in 2012 in comparison to 116,733 in 2002. Direct-to-consumer sales include farmers markets, community supported agriculture (CSA), on-farm stores, and roadside farm stands. Some organic farms also sell products direct to retailer, direct to restaurant and direct to institution. According to the 2008 Organic Production Survey, approximately 7% of organic farm sales went direct-to-consumers, 10% went direct to retailers, and approximately 83% went into wholesale markets. In comparison, only 0.4% of the value of convention agricultural commodities went direct-to-consumers.

While not all products sold at farmer's markets are certified organic, this direct-to-consumer avenue has become increasingly popular in local food distribution and has grown substantially since 1994. In 2014, there were 8,284 farmer's markets in comparison to 3,706 in 2004 and 1,755 in 1994, most of which are found in populated areas such as the Northeast, Midwest, and West Coast.

Labor and Employment

Organic production is more labor-intensive than conventional production. On the one hand, this increased labor cost is one factor that makes organic food more expensive. On the other hand, the increased need for labor may be seen as an "employment dividend" of organic farming, providing more jobs per unit area than conventional systems. The 2011 UNEP Green Economy Report suggests that "[a]n increase in investment in green agriculture is projected to lead to growth in employment of about 60 per cent compared with current levels" and that "green agriculture investments could create 47 million additional jobs compared with BAU2 over the next 40 years." The UNEP also argues that "[b]y greening agriculture and food distribution, more calories per person per day, more jobs and business opportunities especially in rural areas, and market-access opportunities, especially for developing countries, will be available."

World's Food Security

In 2007 the United Nations Food and Agriculture Organization (FAO) said that organic agriculture often leads to higher prices and hence a better income for farmers, so it should be promoted. However, FAO stressed that by organic farming one could not feed the current mankind, even less the bigger future population. Both data and models showed then that organic farming was far from sufficient. Therefore, chemical fertilizers were needed to avoid hunger. Other analysis by many agribusiness executives, agricultural and ecological scientists, and international agriculture experts revealed the opinion that organic farming would not only increase the world's food supply, but might be the only way to eradicate hunger.

FAO stressed that fertilizers and other chemical inputs can much increase the production, particularly in Africa where fertilizers are currently used 90% less than in Asia. For example, in Malawi the yield has been boosted using seeds and fertilizers. FAO also calls for using biotechnology, as it can help smallholder farmers to improve their income and food security.

Also NEPAD, development organization of African governments, announced that feeding Africans and preventing malnutrition requires fertilizers and enhanced seeds.

According to a more recent study in ScienceDigest, organic best management practices shows an average yield only 13% less than conventional. In the world's poorer nations where most of the world's hungry live, and where conventional agriculture's expensive inputs are not affordable by the majority of farmers, adopting organic management actually increases yields 93% on average, and could be an important part of increased food security.

Capacity Building in Developing Countries

Organic agriculture can contribute to ecologically sustainable, socio-economic development, especially in poorer countries. The application of organic principles enables employment of local resources (e.g., local seed varieties, manure, etc.) and therefore cost-effectiveness. Local and international markets for organic products show tremendous growth prospects and offer creative producers and exporters excellent opportunities to improve their income and living conditions.

Organic agriculture is knowledge intensive. Globally, capacity building efforts are underway, including localized training material, to limited effect. As of 2007, the International Federation of Organic Agriculture Movements hosted more than 170 free manuals and 75 training opportunities online.

In 2008 the United Nations Environmental Programme (UNEP) and the United Nations Conference on Trade and Development (UNCTAD) stated that "organic agriculture can be more conducive to food security in Africa than most conventional production systems, and that it is more likely to be sustainable in the long-term" and that "yields had more than doubled where organic, or near-organic practices had been used" and that soil fertility and drought resistance improved.

Millennium Development Goals

The value of organic agriculture (OA) in the achievement of the Millennium Development Goals (MDG), particularly in poverty reduction efforts in the face of climate change, is shown by its contribution to both income and non-income aspects of the MDGs. These benefits are expected to continue in the post-MDG era. A series of case studies conducted in selected areas in Asian countries by the Asian Development Bank Institute (ADBI) and published as a book compilation by ADB in Manila document these contributions to both income and non-income aspects of the MDGs. These include poverty alleviation by way of higher incomes, improved farmers' health owing to less chemical exposure, integration of sustainable principles into rural development policies, improvement of access to safe water and sanitation, and expansion of global partnership for development as small farmers are integrated in value chains.

A related ADBI study also sheds on the costs of OA programs and set them in the context of the costs of attaining the MDGs. The results show considerable variation across the case studies, suggesting that there is no clear structure to the costs of adopting OA. Costs depend on the efficiency of the OA adoption programs. The lowest cost programs were more than ten times less expensive than the highest cost ones. However, further analysis of the gains resulting from OA adoption reveals that the costs per person taken out of poverty was much lower than the estimates of the World Bank, based on income growth in general or based on the detailed costs of meeting some of the more quantifiable MDGs (e.g., education, health, and environment).

Externalities

Agriculture imposes negative externalities (uncompensated costs) upon society through public land and other public resource use, biodiversity loss, erosion, pesticides, nutrient runoff, subsidized water usage, subsidy payments and assorted other problems. Positive externalities include self-reliance, entrepreneurship, respect for nature, and air quality. Organic methods reduce some of these costs. In 2000 uncompensated costs for 1996 reached 2,343 million British pounds or £208 per ha (£84.20/ac). A study of practices in the USA published in 2005 concluded that cropland costs the economy approximately 5 to 16 billion dollars ($30–96/ha - $12–39/ac), while livestock production costs 714 million dollars. Both studies recommended reducing externalities. The 2000 review included reported pesticide poisonings but did not include speculative chronic health effects of pesticides, and the 2004 review relied on a 1992 estimate of the total impact of pesticides.

It has been proposed that organic agriculture can reduce the level of some negative externalities from (conventional) agriculture. Whether the benefits are private or public depends upon the division of property rights.

Several surveys and studies have attempted to examine and compare conventional and organic systems of farming and have found that organic techniques, while not without harm, are less damaging than conventional ones because they reduce levels of biodiversity less than conventional systems do and use less energy and produce less waste when calculated per unit area.

A 2003 to 2005 investigation by the Cranfield University for the Department for Environment Food and Rural Affairs in the UK found that it is difficult to compare the Global Warming Potential (GWP), acidification and eutrophication emissions but "Organic production often results in increased burdens, from factors such as N leaching and N2O emissions", even though primary energy use was less for most organic products. N_2O is always the largest GWP contributor except in tomatoes. However, "organic tomatoes always incur more burdens (except pesticide use)". Some emissions were lower "per area", but organic farming always required 65 to 200% more field area than non-organic farming. The numbers were highest for bread wheat (200+ % more) and potatoes (160% more).

The situation was shown dramatically in a comparison of a modern dairy farm in Wisconsin with one in New Zealand in which the animals grazed extensively. Using total farm emissions per kg milk produced as a parameter, the researchers showed that production of methane from belching was higher in the New Zealand farm, while carbon dioxide production was higher in the Wisconsin farm. Output of nitrous oxide, a gas with an estimated global warming potential 310 times that of carbon dioxide was also higher in the New Zealand farm. Methane from manure handling was similar in the two types of farm. The explanation for the finding relates to the different diets used on these farms, being based more completely on forage (and hence more fibrous) in New Zealand and containing less concentrate than in Wisconsin. Fibrous diets promote a higher proportion of acetate in the gut of ruminant animals, resulting in a higher production of methane that must be released by belching. When cattle are given a diet containing some concentrates (such as corn and soybean meal) in addition to grass and silage, the pattern of ruminal fermentation alters from acetate to mainly propionate. As a result, methane production is reduced. Capper et al. compared the environmental impact of US dairy production in 1944 and 2007. They calculated that the carbon "footprint" per billion kg (2.2 billion lb) of milk produced in 2007 was 37 percent that of equivalent milk production in 1944.

Environmental Impact and Emissions

Researchers at Oxford university analyzed 71 peer-reviewed studies and observed that organic products are sometimes worse for the environment. Organic milk, cereals, and pork generated higher greenhouse gas emissions per product than conventional ones but organic beef and olives had lower emissions in most studies. Usually organic products required less energy, but more land. Per unit of product, organic produce generates higher nitrogen leaching, nitrous oxide emissions, ammonia emissions, eutrophication and acidification potential than when conventionally grown. Other differences were not significant. The researchers concluded, as there is not singular way of doing conventional or organic farming, that the debate should go beyond the conventional vs organic debate, and more about finding specific solutions to specific circumstances.

Proponents of organic farming have claimed that organic agriculture emphasizes closed nutrient cycles, biodiversity, and effective soil management providing the capacity to mitigate and even reverse the effects of climate change and that organic agriculture can decrease fossil fuel emissions. "The carbon sequestration efficiency of organic systems in temperate climates is almost double

(575-700 kg carbon per ha per year - 510-625 lb/ac/an) that of conventional treatment of soils, mainly owing to the use of grass clovers for feed and of cover crops in organic rotations."

Critics of organic farming methods believe that the increased land needed to farm organic food could potentially destroy the rainforests and wipe out many ecosystems.

Nutrient Leaching

According to the meta-analysis of 71 studies, nitrogen leaching, nitrous oxide emissions, ammonia emissions, eutrophication potential and acidification potential were higher for organic products, although in one study "nitrate leaching was 4.4-5.6 times higher in conventional plots than organic plots".

Excess nutrients in lakes, rivers, and groundwater can cause algal blooms, eutrophication, and subsequent dead zones. In addition, nitrates are harmful to aquatic organisms by themselves.

Land Use

The Oxford meta-analysis of 71 studies proved that organic farming requires 84% more land, mainly due to lack of nutrients but sometimes due to weeds, diseases or pests, lower yielding animals and land required for fertility building crops. While organic farming does not necessarily save land for wildlife habitats and forestry in all cases, the most modern breakthroughs in organic are addressing these issues with success.

Professor Wolfgang Branscheid says that organic animal production is not good for the environment, because organic chicken requires doubly as much land as "conventional" chicken and organic pork a quarter more. According to a calculation by Hudson Institute, organic beef requires triply as much land. On the other hand, certain organic methods of animal husbandry have been shown to restore desertified, marginal, and/or otherwise unavailable land to agricultural productivity and wildlife. Or by getting both forage and cash crop production from the same fields simultaneously, reduce net land use.

In England organic farming yields 55% of normal yields. While in other regions of the world, organic methods have started producing record yields.

Pesticides

A sign outside of an organic apple orchard in Pateros, Washington reminding orchardists not to spray pesticides on these trees.

In organic farming synthetic pesticides are generally prohibited. A chemical is said to be synthetic if it does not already exist in the natural world. But the organic label goes further and usually prohibit compounds that exist in nature if they are produced by Chemical synthesis. So the prohibition is also about the method of production and not only the nature of the compound.

An non exhaustive list of organic approved pesticides with theirs Median lethal dose

- Copper(II) sulfate is used as a fungicide and is also used in conventional agriculture (LD_{50} 300 mg/kg). Conventional agriculture has the option to use the less toxic Mancozeb (LD_{50} 4,500 to 11,200 mg/kg)

- Boric acid is used as stomach poison that target insects (LD_{50}: 2660 mg/kg).

- Pyrethrin comes from chemicals extracted from flowers of the genus Pyrethrum (LD_{50} of 370 mg/kg). Its potent toxicity is used to control insects.

- Lime sulphur (aka calcium polysulfide) and sulfur are considered to be allowed, synthetic

materials (LD_{50}: 820 mg/kg)

- Rotenone is a powerful insecticide that was used to control insects (LD_{50}: 132 mg/kg). Despite the high toxicity of Rotenone to aquatic life and some links to Parkinson disease the compound is still allowed in organic farming as it is a naturally occurring compound.

- Bromomethane is a gas that is still used in the nurseries of Strawberry organic farming

- Azadirachtin is a wide spectrum very potent insecticide. Almost non toxic to mammals (LD_{50} in rats is > 3,540 mg/kg) but affects beneficial insects.

Food Quality and Safety

While there may be some differences in the amounts of nutrients and anti-nutrients when organically produced food and conventionally produced food are compared, the variable nature of food production and handling makes it difficult to generalize results, and there is insufficient evidence to make claims that organic food is safer or healthier than conventional food. Claims that organic food tastes better are not supported by evidence.

Soil Conservation

Supporters claim that organically managed soil has a higher quality and higher water retention. This may help increase yields for organic farms in drought years. Organic farming can build up soil organic matter better than conventional no-till farming, which suggests long-term yield benefits from organic farming. An 18-year study of organic methods on nutrient-depleted soil concluded that conventional methods were superior for soil fertility and yield for nutrient-depleted soils in cold-temperate climates, arguing that much of the benefit from organic farming derives from imported materials that could not be regarded as self-sustaining.

In *Dirt: The Erosion of Civilizations*, geomorphologist David Montgomery outlines a coming crisis from soil erosion. Agriculture relies on roughly one meter of topsoil, and that is being depleted ten times faster than it is being replaced. No-till farming, which some claim depends upon pesticides, is one way to minimize erosion. However, a 2007 study by the USDA's Agricultural Research Service has found that manure applications in tilled organic farming are better at building up the soil than no-till.

Biodiversity

The conservation of natural resources and biodiversity is a core principle of organic production. Three broad management practices (prohibition/reduced use of chemical pesticides and inorganic fertilizers; sympathetic management of non-cropped habitats; and preservation of mixed farming) that are largely intrinsic (but not exclusive) to organic farming are particularly beneficial for farmland wildlife. Using practices that attract or introduce beneficial insects, provide habitat for birds and mammals, and provide conditions that increase soil biotic diversity serve to supply vital ecological services to organic production systems. Advantages to certified organic operations that implement these types of production practices include: 1) decreased dependence on outside fertility inputs; 2) reduced pest management costs; 3) more reliable sources of clean water; and 4) better pollination.

Nearly all non-crop, naturally occurring species observed in comparative farm land practice studies show a preference for organic farming both by abundance and diversity. An average of 30% more species inhabit organic farms. Birds, butterflies, soil microbes, beetles, earthworms, spiders, vegetation, and mammals are particularly affected. Lack of herbicides and pesticides improve biodiversity fitness and population density. Many weed species attract beneficial insects that improve soil qualities and forage on weed pests. Soil-bound organisms often benefit because of increased bacteria populations due to natural fertilizer such as manure, while experiencing reduced intake of herbicides and pesticides. Increased biodiversity, especially from beneficial soil microbes and mycorrhizae have been proposed as an explanation for the high yields experienced by some organic plots, especially in light of the differences seen in a 21-year comparison of organic and control fields.

Biodiversity from organic farming provides capital to humans. Species found in organic farms enhance sustainability by reducing human input (e.g., fertilizers, pesticides).

The USDA's Agricultural Marketing Service (AMS) published a *Federal Register* notice on 15 January 2016, announcing the National Organic Program (NOP) final guidance on Natural Resources and Biodiversity Conservation for Certified Organic Operations. Given the broad scope of natural resources which includes soil, water, wetland, woodland and wildlife, the guidance provides examples of practices that support the underlying conservation principles and demonstrate compliance with USDA organic regulations § 205.200. The final guidance provides organic certifiers and farms with examples of production practices that support conservation principles and comply with the USDA organic regulations, which require operations to maintain or improve natural resources. The final guidance also clarifies the role of certified operations (to submit an OSP to a certifier), certifiers (ensure that the OSP describes or lists practices that explain the operator's monitoring plan and practices to support natural resources and biodiversity conservation), and inspectors (onsite inspection) in the implementation and verification of these production practices.

A wide range of organisms benefit from organic farming, but it is unclear whether organic methods confer greater benefits than conventional integrated agri-environmental programs. Organic farming is often presented as a more biodiversity-friendly practice, but the generality of the beneficial effects of organic farming is debated as the effects appear often species- and context-dependent, and current research has highlighted the need to quantify the relative effects of local- and landscape-scale management on farmland biodiversity. There are four key issues when comparing the impacts on biodiversity of organic and conventional farming: (1) It remains unclear whether a holistic whole-farm approach (i.e. organic) provides greater benefits to biodiversity than carefully targeted prescriptions applied to relatively small areas of cropped and/or non-cropped habitats within conventional agriculture (i.e. agri-environment schemes); (2) Many comparative studies encounter methodological problems, limiting their ability to draw quantitative conclusions; (3) Our knowledge of the impacts of organic farming in pastoral and upland agriculture is limited; (4) There remains a pressing need for longitudinal, system-level studies in order to address these issues and to fill in the gaps in our knowledge of the impacts of organic farming, before a full appraisal of its potential role in biodiversity conservation in agroecosystems can be made.

Regional Support for Organic Farming

India

In India, states such as Sikkim and Kerala have planned to shift to fully organic cultivation by 2015 and 2016 respectively.

Subsistence Agriculture

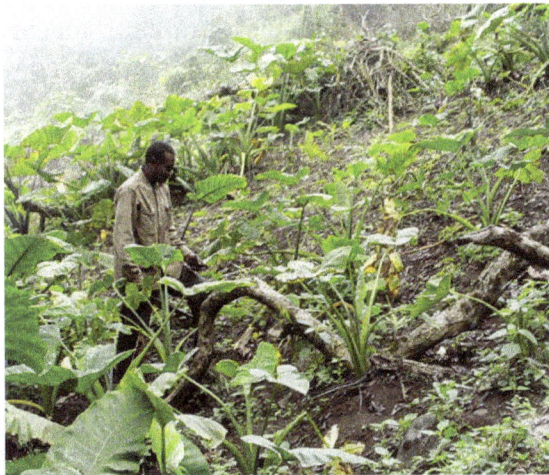

Like most farmers in Sub-Saharan Africa, this Cameroonian man cultivates at the subsistence level.

Subsistence agriculture is self-sufficiency farming in which the farmers focus on growing enough food to feed themselves and their families. The output is mostly for local requirements with little or no surplus for trade. The typical subsistence farm has a range of crops and animals needed by the family to feed and clothe themselves during the year. Planting decisions are made principally with an eye toward what the family will need during the coming year, and secondarily toward market prices. Tony Waters writes: "Subsistence peasants are people who grow what they eat, build their own houses, and live without regularly making purchases in the marketplace."

However, despite the primacy of self-sufficiency in subsistence farming, today most subsistence farmers also participate in trade to some degree, though usually it is for goods that are not necessary for survival, and may include sugar, iron roofing sheets, bicycles, used clothing, and so forth. Most subsistence farmers today reside in developing countries, although their amount of trade as measured in cash is less than that of consumers in countries with modern complex markets, many have important trade contacts and trade items that they can produce because of their special skills or special access to resources valued in the marketplace.

Subsistence agriculture also emerged independently in Mexico where it was based on maize cultivation, and the Andes where it was based on the domestication of the potato. Subsistence agriculture was the dominant mode of production in the world until recently, when market-based capitalism became widespread. Subsistence horticulture may have developed independently in South East Asia and Papua New Guinea.

Subsistence farming continues today in large parts of rural Africa, and parts of Asia and Latin America. Subsistence agriculture had largely disappeared in Europe by the beginning of World War I, and in North America with the movement of sharecroppers and tenant farmers out of the American South and Midwest during the 1930s and 1940s. As recently as the 1950s, it was still common on family farms in North America and Europe to grow much of a family's own food and make much of its own clothing, although sales of some of the farm's production earned enough currency to buy certain staples, typically including sugar; coffee and tea; petroleum distillates (petrol, kerosene, fuel oil); textile products such as bolts of cloth, needles, and thread; medicines; hardware products such as nails, screws, and wire; and a few discretionary items such as candy or books. Many of the preceding items, as well as occasional services from physicians, veterinarians, blacksmiths, and others, were often bought with barter rather than currency. In Central and Eastern Europe subsistence and semi-subsistence agriculture reappeared within the transition economy since about 1990.

Types of Subsistence Farming

Shifting Agriculture

In this type of agriculture, a patch of forest land is cleared by a combination of felling and burning, and crops are grown. After 2-3 years the fertility of the soil begins to decline, the land is abandoned and the farmer moves to clear a fresh piece of land elsewhere in the forest as the process continues. While the land is left fallow the forest regrows in the cleared area and soil fertility and biomass is restored. After a decade or more, the farmer may return to the first piece of land. This form of agriculture is sustainable at low population densities, but higher population loads require more frequent clearing which prevents soil fertility from recovering, opens up more of the forest canopy, and encourages scrub at the expense of large trees, eventually resulting in deforestation and land erosion.

Primitive Agriculture

While this 'slash-and-burn' technique may describe the method for opening new land, commonly thc farmcrs in qucstion havc in cxistence at the same time smaller fields, sometimes merely gardens, near the homestead there they practice intensive 'non-shifting" techniques until shortage of fields where they can employ "slash and burn" to clear land and (by the burning) provide fertilizer (ash). Such gardens nearer the homestead often regularly receive household refuse, the manure of any household chickens or goats, and compost piles where refuse is thrown initially just to get it out of the way. However, such farmers often recognize the value of such compost and apply it regularly to their smaller fields. They also may irrigate part of such fields if they are near a source of water.

In some areas of tropical Africa, at least, such smaller fields may be ones in which crops are grown on raised beds. Thus farmers practicing 'slash and burn' agriculture are often much more sophisticated agriculturalists than the term "slash and burn" subsistence farmers suggest.

Nomadic Herding

In this type of farming people migrate along with their animals from one place to another in search of fodder for their animals. Generally they rear cattle, sheep, goats, camels and/or yaks for milk, skin, meat and wool. This way of life is common in parts of central and western Asia, India, east and south-west Africa and northern Eurasia. Examples are the nomadic Bhotiyas and Gujjars of the Himalayas.

Intensive Subsistence Farming

In Intensive subsistence agriculture, the farmer cultivates a small plot of land using simple tools and more labour. Climate, with large number of days with sunshine and fertile soils permits growing of more than one crop annually on the same plot. Farmers use their small land holdings to produce enough, for their local consumption, while remaining produce is used for exchange against other goods. It results in much more food being produced per acre compared to other subsistence patterns. In the most intensive situation, farmers may even create terraces along steep hillsides to cultivate rice paddies. Such fields are found in densely populated parts of Asia, such as in The Philippines. They may also intensify by using manure, artificial irrigation and animal waste as fertilizer. Intensive subsistence farming is prevalent in the thickly populated areas of the monsoon regions of south, southwest, and east Asia.

Intensive Farming

Intensive farming or intensive agriculture is any of various types of agriculture that involve higher levels of input and output per unit of agricultural land area. It is characterized by a low fallow ratio, higher use of inputs such as capital and labour, and higher crop yields per unit land area. This is in contrast to traditional agriculture in which the inputs per unit land are lower.The term "intensive" has various senses, some of which refer to organic farming methods (such as biointensive agriculture and French intensive gardening) and others of which refer to nonorganic and industrial methods. Intensive animal farming involves either large numbers of animals raised on limited land, usually confined animal feeding operations (CAFO) often referred to as factory farms, or managed intensive rotational grazing (MIRG), which has both organic and nonorganic types. Both increase the yields of food and fiber per acre as compared to traditional animal husbandry. In a CAFO feed is brought to the animals, which are seldom moved, while in MIRG the animals are repeatedly moved to fresh forage.

Industrial agriculture is characterised by innovations designed to increase yield. Techniques include planting multiple crops per year, reducing the frequency of fallow years and improving cultivars. It also involves increased use of fertilizers, plant growth regulators, pesticides and mechanization, controlled by increased and more detailed analysis of growing conditions, including weather, soil, water, weeds and pests. This system is supported by ongoing innovation in agricultural machinery and farming methods, genetic technology, techniques for achieving economies of scale, logistics and data collection and analysis technology. Intensive farms are widespread in developed nations and increasingly prevalent worldwide. Most of the meat, dairy, eggs, fruits and vegetables available in supermarkets are produced by such farms.

Smaller intensive farms usually include higher inputs of labor and more often use sustainable intensive methods. The farming practices commonly found on such farms are referred to as appropriate technology. These farms are less widespread in both developed countries and worldwide, but are growing more rapidly. Most of the food available in specialty markets such as farmers markets is produced by these smallholder farms.

History

Early 20th-century image of a tractor ploughing an alfalfa field

Agricultural development in Britain between the 16th century and the mid-19th century saw a massive increase in agricultural productivity and net output. This in turn supported unprecedented population growth, freeing up a significant percentage of the workforce, and thereby helped enable the Industrial Revolution. Historians cited enclosure, mechanization, four-field crop rotation, and selective breeding as the most important innovations.

Industrial agriculture arose along with the Industrial Revolution. By the early 19th century, agricultural techniques, implements, seed stocks and cultivars had so improved that yield per land unit was many times that seen in the Middle Ages.

The industrialization phase involved a continuing process of mechanization. Horse-drawn machinery such as the McCormick reaper revolutionized harvesting, while inventions such as the cotton gin reduced the cost of processing. During this same period, farmers began to use steam-powered threshers and tractors, although they were expensive and dangerous. In 1892, the first gasoline-powered tractor was successfully developed, and in 1923, the International Harvester Farmall tractor became the first all-purpose tractor, marking an inflection point in the replacement of draft animals with machines. Mechanical harvesters (combines), planters, transplanters and other equipment were then developed, further revolutionizing agriculture. These inventions increased yields and allowed individual farmers to manage increasingly large farms.

The identification of nitrogen, potassium, and phosphorus (NPK) as critical factors in plant growth led to the manufacture of synthetic fertilizers, further increasing crop yields. In 1909 the Haber-Bosch method to synthesize ammonium nitrate was first demonstrated. NPK fertilizers stimulated the first concerns about industrial agriculture, due to concerns that they came with serious side effects such as soil compaction, soil erosion and declines in overall soil fertility, along with health concerns about toxic chemicals entering the food supply.

The identification of carbon as a critical factor in plant growth and soil health, particularly in the form of humus, led to so-called *sustainable agriculture*, alternative forms of intensive agriculture that also surpass traditional agriculture, without side effects or health issues. Farmers adopting this approach were initially referred to as *humus farmers*, later as *organic farmers*.

The discovery of vitamins and their role in nutrition, in the first two decades of the 20th century, led to vitamin supplements, which in the 1920s allowed some livestock to be raised indoors, reducing their exposure to adverse natural elements. Chemicals developed for use in World War II gave rise to synthetic pesticides.

Following World War II, synthetic fertilizer use increased rapidly, while sustainable intensive farming advanced much more slowly. Most of the resources in developed nations went to improving industrial intensive farming, and very little went to improving organic farming. Thus, particularly in the developed nations, industrial intensive farming grew to become the dominant form of agriculture.

The discovery of antibiotics and vaccines facilitated raising livestock in CAFOs by reducing diseases caused by crowding. Developments in logistics and refrigeration as well as processing technology made long-distance distribution feasible.

Between 1700 and 1980, "the total area of cultivated land worldwide increased 466%" and yields increased dramatically, particularly because of selectively bred high-yielding varieties, fertilizers, pesticides, irrigation and machinery. Global agricultural production doubled between 1820 1920; between 1920 and 1950; between 1950 and 1965; and again between 1965 and 1975 to feed a global population that grew from one billion in 1800 to 6.5 billion in 2002. The number of people involved in farming in industrial countries dropped, from 24 percent of the American population to 1.5 percent in 2002. In 1940, each farmworker supplied 11 consumers, whereas in 2002, each worker supplied 90 consumers. The number of farms also decreased and their ownership became more concentrated. In 2000 in the U.S., four companies produced 81 percent of cows, 73 percent of sheep, 57 percent of pigs, and 50 percent of chickens, cited as an example of "vertical integration" by the president of the U.S. National Farmers' Union. Between 1967 and 2002 the one million pig farms in America consolidated into 114,000 with 80 million pigs (out of 95 million) produced each year on factory farms, according to the U.S. National Pork Producers Council. According to the

Worldwatch Institute, 74 percent of the world's poultry, 43 percent of beef, and 68 percent of eggs are produced this way.

Concerns over the sustainability of industrial agriculture, which has become associated with decreased soil quality, and over the environmental effects of fertilizers and pesticides, have not subsided. Alternatives such as integrated pest management (IPM) have had little impact because policies encourage the use of pesticides and IPM is knowledge-intensive. These concerns sustained the organic movement and caused a resurgence in sustainable intensive farming and funding for the development of appropriate technology.

Famines continued throughout the 20th century. Through the effects of climactic events, government policy, war and crop failure, millions of people died in each of at least ten famines between the 1920s and the 1990s.

Techniques and Technologies

Livestock

A commercial chicken house raising broiler pullets for meat.

Confined Animal Feeding Operations

Intensive livestock farming, also called "factory farming" is a term referring to the process of raising livestock in confinement at high stocking density. "Concentrated animal feeding operations" (CAFO) or "intensive livestock operations", can hold large numbers (some up to hundreds of thousands) of cows, hogs, turkeys or chickens, often indoors. The essence of such farms is the concentration of livestock in a given space. The aim is to provide maximum output at the lowest possible cost and with the greatest level of food safety. The term is often used pejoratively. However, CAFOs have dramatically increased the production of food from animal husbandry worldwide, both in terms of total food produced and efficiency.

Food and water is delivered to the animals, and therapeutic use of antimicrobial agents, vitamin supplements and growth hormones are often employed. Growth hormones are not used on chickens nor on any animal in the European Union. Undesirable behaviours often related to the stress of confinement led to a search for docile breeds (e.g., with natural dominance behaviours bred out), physical restraints to stop interaction, such as individual cages for chickens, or physically modification such as the de-beaking of chickens to reduce the harm of fighting.

The CAFO designation resulted from the 1972 US Federal Clean Water Act, which was enacted to protect and restore lakes and rivers to a "fishable, swimmable" quality. The United States Environmental Protection Agency (EPA) identified certain animal feeding operations, along with many other types of industry, as "point source" groundwater polluters. These operations were subjected to regulation.

In 17 states in the U.S., isolated cases of groundwater contamination were linked to CAFOs. For example, the ten million hogs in North Carolina generate 19 million tons of waste per year. The U.S. federal government acknowledges the waste disposal issue and requires that animal waste be stored in lagoons. These lagoons can be as large as 7.5 acres (30,000 m²). Lagoons not protected with an impermeable liner can leak into groundwater under some conditions, as can runoff from manure used as fertilizer. A lagoon that burst in 1995 released 25 million gallons of nitrous sludge in North Carolina's New River. The spill allegedly killed eight to ten million fish.

The large concentration of animals, animal waste and dead animals in a small space poses ethical issues to some consumers. Animal rights and animal welfare activists have charged that intensive animal rearing is cruel to animals.

Other concerns include persistent noxious odor, the effects on human health and the role of antibiotics use in the rise of resistant infectious bacteria.

According to the U.S. Centers for Disease Control and Prevention (CDC), farms on which animals are intensively reared can cause adverse health reactions in farm workers. Workers may develop acute and/or chronic lung disease, musculoskeletal injuries and may catch (zoonotic) infections from the animals.

Managed Intensive Rotational Grazing

Managed Intensive Rotational Grazing (MIRG), also known as cell grazing, mob grazing and holistic managed planned grazing, is a variety of forage use in which herds/flocks are regularly and systematically moved to fresh, rested grazing areas to maximize the quality and quantity of forage growth. MIRG can be used with cattle, sheep, goats, pigs, chickens, turkeys, ducks and other animals. The herds graze one portion of pasture, or a paddock, while allowing the others to recover. Resting grazed lands allows the vegetation to renew energy reserves, rebuild shoot systems, and deepen root systems, resulting in long-term maximum biomass production. MIRG is especially effective because grazers thrive on the more tender younger plant stems. MIRG also leave parasites behind to die off minimizing or eliminating the need for de-wormers. Pasture systems alone can allow grazers to meet their energy requirements, and with the increased productivity of MIRG systems, the animals obtain the majority of their nutritional needs, in some cases all, without the supplemental feed sources that are required in continuous grazing systems or CAFOs.

Crops

The Green Revolution transformed farming in many developing countries. It spread technologies that had already existed, but had not been widely used outside of industrialized nations. These technologies included "miracle seeds", pesticides, irrigation and synthetic nitrogen fertilizer.

Seeds

In the 1970s scientists created strains of maize, wheat, and rice that are generally referred to as high-yielding varieties (HYV). HYVs have an increased nitrogen-absorbing potential compared to other varieties. Since cereals that absorbed extra nitrogen would typically lodge (fall over) before harvest, semi-dwarfing genes were bred into their genomes. Norin 10 wheat, a variety developed by Orville Vogel from Japanese dwarf wheat varieties, was instrumental in developing wheat cultivars. IR8, the first widely implemented HYV rice to be developed by the International Rice Research Institute, was created through a cross between an Indonesian variety named "Peta" and a Chinese variety named "Dee Geo Woo Gen."

With the availability of molecular genetics in Arabidopsis and rice the mutant genes responsible (*reduced height (rht), gibberellin insensitive (gai1)* and *slender rice (slr1)*) have been cloned and identified as cellular signalling components of gibberellic acid, a phytohormone involved in regulating stem growth via its effect on cell division. Photosynthetic investment in the stem is reduced dramatically as the shorter plants are inherently more mechanically stable. Nutrients become redirected to grain production, amplifying in particular the yield effect of chemical fertilisers.

HYVs significantly outperform traditional varieties in the presence of adequate irrigation, pesticides and fertilizers. In the absence of these inputs, traditional varieties may outperform HYVs. They were developed as F1 hybrids, meaning seeds need to be purchased every season to obtain maximum benefit, thus increasing costs.

Crop Rotation

Crop rotation or crop sequencing is the practice of growing a series of dissimilar types of crops in the same space in sequential seasons for benefits such as avoiding pathogen and pest buildup that occurs when one species is continuously cropped. Crop rotation also seeks to balance the nutrient demands of various crops to avoid soil nutrient depletion. A traditional component of crop rotation is the replenishment of nitrogen through the use of legumes and green manure in sequence with cereals and other crops. Crop rotation can also improve soil structure and fertility by alternating deep-rooted and shallow-rooted plants. One technique is to plant multi-species cover crops between commercial crops. This combines the advantages of intensive farming with continuous cover and polyculture.

Irrigation

Overhead irrigation, center pivot designed

Crop irrigation accounts for 70% of the world's fresh water use.

Flood irrigation, the oldest and most common type, is typically unevenly distributed, as parts of a field may receive excess water in order to deliver sufficient quantities to other parts. Overhead irrigation, using center-pivot or lateral-moving sprinklers, gives a much more equal and controlled distribution pattern. Drip irrigation is the most expensive and least-used type, but delivers water to plant roots with minimal losses.

Water catchment management measures include recharge pits, which capture rainwater and runoff and use it to recharge groundwater supplies. This helps in the replenishment of groundwater wells and eventually reduces soil erosion. Dammed rivers creating Reservoirs store water for irrigation and other uses over large areas. Smaller areas sometimes use irrigation ponds or groundwater.

Weed Control

In agriculture, systematic weed management is usually required, often performed by machines such as cultivators or liquid herbicide sprayers. Herbicides kill specific targets while leaving the crop relatively unharmed. Some of these act by interfering with the growth of the weed and are often based on plant hormones. Weed control through herbicide is made more difficult when the weeds become resistant to the herbicide. Solutions include:

- Cover crops (especially those with allelopathic properties) that out-compete weeds or inhibit their regeneration.

- Multiple herbicides, in combination or in rotation

- Strains genetically engineered for herbicide tolerance

- Locally adapted strains that tolerate or out-compete weeds

- Tilling

- Ground cover such as mulch or plastic

- Manual removal

- Mowing

- Grazing

- Burning

Terracing

In agriculture, a terrace is a leveled section of a hilly cultivated area, designed as a method of soil conservation to slow or prevent the rapid surface runoff of irrigation water. Often such land is formed into multiple terraces, giving a stepped appearance. The human landscapes of rice cultivation in terraces that follow the natural contours of the escarpments like contour ploughing is a classic feature of the island of Bali and the Banaue Rice Terraces in Banaue, Ifugao, Philippines. In Peru, the Inca made use of otherwise unusable slopes by drystone walling to create terraces.

Terrace rice fields in Yunnan Province, China

Rice Paddies

A paddy field is a flooded parcel of arable land used for growing rice and other semiaquatic crops. Paddy fields are a typical feature of rice-growing countries of east and southeast Asia including Malaysia, China, Sri Lanka, Myanmar, Thailand, Korea, Japan, Vietnam, Taiwan, Indonesia, India, and the Philippines. They are also found in other rice-growing regions such as Piedmont (Italy), the Camargue (France) and the Artibonite Valley (Haiti). They can occur naturally along rivers or marshes, or can be constructed, even on hillsides. They require large water quantities for irrigation, much of it from flooding. It gives an environment favourable to the strain of rice being grown, and is hostile to many species of weeds. As the only draft animal species which is comfortable in wetlands, the water buffalo is in widespread use in Asian rice paddies.

Paddy-based rice-farming has been practiced in Korea since ancient times. A pit-house at the Dae-cheon-ni archaeological site yielded carbonized rice grains and radiocarbon dates indicating that rice cultivation may have begun as early as the Middle Jeulmun Pottery Period (c. 3500-2000 BC) in the Korean Peninsula. The earliest rice cultivation there may have used dry-fields instead of paddies.

The earliest Mumun features were usually located in naturally swampy, low-lying narrow gulleys and fed by local streams. Some Mumun paddies in flat areas were made of a series of squares and rectangles separated by bunds approximately 10 cm in height, while terraced paddies consisted of long irregularly shapes that followed natural contours of the land at various levels.

Like today's, Mumun period rice farmers used terracing, bunds, canals and small reservoirs. Some paddy-farming techniques of the Middle Mumun (c. 850-550 BC) can be interpreted from the well-preserved wooden tools excavated from archaeological rice paddies at the Majeon-ni Site. However, iron tools for paddy-farming were not introduced until sometime after 200 BC. The spatial scale of individual paddies, and thus entire paddy-fields, increased with the regular use of iron tools in the Three Kingdoms of Korea Period (c. AD 300/400-668).

A recent development in the intensive production of rice is System of Rice Intensification (SRI). Developed in 1983 by the French Jesuit Father Henri de Laulanié in Madagascar, by 2013 the number of smallholder farmers using SRI had grown to between 4 and 5 million.

Aquaculture

Aquaculture is the cultivation of the natural products of water (fish, shellfish, algae, seaweed and other aquatic organisms). Intensive aquaculture takes place on land using tanks, ponds or other controlled systems or in the ocean, using cages.

Sustainable Intensive Farming

Sustainable intensive farming practises have been developed to slow the deterioration of agricultural land and even regenerate soil health and ecosystem services, while still offering high yields. Most of these developments fall in the category of organic farming, or the integration of organic and conventional agriculture.

"Organic systems and the practices that make them effective are being picked up more and more by conventional agriculture and will become the foundation for future farming systems. They won't be called organic, because they'll still use some chemicals and still use some fertilizers, but they'll function much more like today's organic systems than today's conventional systems."

Dr. Charles Benbrook Executive director US House Agriculture Subcommittee Director Agricultural Board - National Academy Sciences (FMR)

The System of Crop Intensification (SCI) was born out of research primarily at Cornell University and smallholder farms in India on SRI. It uses the SRI concepts and methods for rice and applies them to crops like wheat, sugarcane, finger millet, and others. It can be 100% organic, or integrated with reduced conventional inputs.

Holistic management is a systems thinking approach that was originally developed for reversing desertification. Holistic planned grazing is similar to rotational grazing but differs in that it more explicitly provides a framework for adapting to four basic ecosystem processes: the water cycle, the mineral cycle including the carbon cycle, energy flow and community dynamics (the relationship between organisms in an ecosystem) as equal in importance to livestock production and social welfare. By intensively managing the behavior and movement of livestock, holistic planned grazing simultaneously increases stocking rates and restores grazing land.

Pasture cropping plants grain crops directly into grassland without first applying herbicides. The perennial grasses form a living mulch understory to the grain crop, eliminating the need to plant cover crops after harvest. The pasture is intensively grazed both before and after grain production using holistic planned grazing. This intensive system yields equivalent farmer profits (partly from increased livestock forage) while building new topsoil and sequestering up to 33 tons of CO_2/ha/year.

The Twelve Aprils grazing program for dairy production, developed in partnership with US-DA-SARE, is similar to pasture cropping, but the crops planted into the perennial pasture are forage crops for dairy herds. This system improves milk production and is more sustainable than confinement dairy production.

Integrated Multi-Trophic Aquaculture (IMTA) is an example of a holistic approach. IMTA is a practice in which the by-products (wastes) from one species are recycled to become inputs (fertilizers, food) for another. Fed aquaculture (e.g. fish, shrimp) is combined with inorganic extractive (e.g. seaweed) and organic extractive (e.g. shellfish) aquaculture to create balanced systems for environmental sustainability (biomitigation), economic stability (product diversification and risk reduction) and social acceptability (better management practices).

Biointensive agriculture focuses on maximizing efficiency such as per unit area, energy input and water input. Agroforestry combines agriculture and orchard/forestry technologies to create more integrated, diverse, productive, profitable, healthy and sustainable land-use systems.

Intercropping can increase yields or reduce inputs and thus represents (potentially sustainable) agricultural intensification. However, while total yield per acre is often increased dramatically, yields of any single crop often diminish. There are also challenges to farmers relying on farming equipment optimized for monoculture, often resulting in increased labor inputs.

Vertical farming is intensive crop production on a large scale in urban centers in multi-story, artificially-lit structures that uses far less inputs and produces fewer environmental impacts.

An integrated farming system is a progressive biologically integrated sustainable agriculture system such as IMTA or Zero waste agriculture whose implementation requires exacting knowledge of the interactions of multiple species and whose benefits include sustainability and increased profitability. Elements of this integration can include:

- Intentionally introducing flowering plants into agricultural ecosystems to increase pollen-and nectar-resources required by natural enemies of insect pests

- Using crop rotation and cover crops to suppress nematodes in potatoes

Challenges

The challenges and issues of industrial agriculture for society, for the industrial agriculture sector, for the individual farm, and for animal rights include the costs and benefits of both current practices and proposed changes to those practices. This is a continuation of thousands of years of invention in feeding ever growing populations.

When hunter-gatherers with growing populations depleted the stocks of game and wild foods across the Near East, they were forced to introduce agriculture. But agriculture brought much longer hours of work and a less rich diet than hunter-gatherers enjoyed. Further population growth among shifting slash-and-burn farmers led to shorter fallow periods, falling yields and soil erosion. Plowing and fertilizers were introduced to deal with these problems - but once again involved longer hours of work and degradation of soil resources(Boserup, The Conditions of Agricultural Growth, Allen and Unwin, 1965, expanded and updated in Population and Technology, Blackwell, 1980.).

While the point of industrial agriculture is to profitably supply the world at the lowest cost, industrial methods have significant side effects. Further, industrial agriculture is not an indivisible whole, but instead is composed of multiple elements, each of which can be modified in response to market conditions, government regulation and further innovation and has its own side-effects. Various interest groups reach different conclusions on the subject.

Benefits

Population Growth

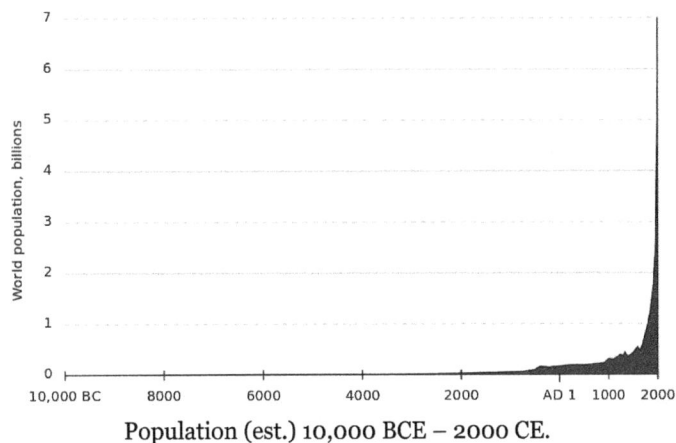

Population (est.) 10,000 BCE – 2000 CE.

Very roughly:

- 30,000 years ago hunter-gatherer behavior fed 6 million people

- 3,000 years ago primitive agriculture fed 60 million people

- 300 years ago intensive agriculture fed 600 million people

- Today industrial agriculture attempts to feed 6 billion people

Estimated world population at various dates, in **thousands**							
Year	World	Africa	Asia	Europe	Central & South America	North America*	Oceania
8000 BCE	8 000						
1000 BCE	50 000						
500 BCE	100 000						
1 CE	200,000 plus						
1000	310 000						
1750	791 000	106 000	502 000	163 000	16 000	2 000	2 000
1800	978 000	107 000	635 000	203 000	24 000	7 000	2 000
1850	1 262 000	111 000	809 000	276 000	38 000	26 000	2 000
1900	1 650 000	133 000	947 000	408 000	74 000	82 000	6 000
1950	2 518 629	221 214	1 398 488	547 403	167 097	171 616	12 812

1955	2 755 823	246 746	1 541 947	575 184	190 797	186 884	14 265
1960	2 981 659	277 398	1 674 336	601 401	209 303	204 152	15 888
1965	3 334 874	313 744	1 899 424	634 026	250 452	219 570	17 657
1970	3 692 492	357 283	2 143 118	655 855	284 856	231 937	19 443
1975	4 068 109	408 160	2 397 512	675 542	321 906	243 425	21 564
1980	4 434 682	469 618	2 632 335	692 431	361 401	256 068	22 828
1985	4 830 979	541 814	2 887 552	706 009	401 469	269 456	24 678
1990	5 263 593	622 443	3 167 807	721 582	441 525	283 549	26 687
1995	5 674 380	707 462	3 430 052	727 405	481 099	299 438	28 924
2000	6 070 581	795 671	3 679 737	727 986	520 229	315 915	31 043
2005	6 453 628	887 964	3 917 508	724 722	558 281	332 156	32 998**

An example of industrial agriculture providing cheap and plentiful food is the U.S.'s "most successful program of agricultural development of any country in the world". Between 1930 and 2000 U.S. agricultural productivity (output divided by all inputs) rose by an average of about 2 percent annually causing food prices to decrease. "The percentage of U.S. disposable income spent on food prepared at home decreased, from 22 percent as late as 1950 to 7 percent by the end of the century."

Liabilities

Environment

Industrial agriculture uses huge amounts of water, energy, and industrial chemicals; increasing pollution in the arable land, usable water and atmosphere. Herbicides, insecticides and fertilizers are accumulating in ground and surface waters. "Many of the negative effects of industrial agriculture are remote from fields and farms. Nitrogen compounds from the Midwest, for example, travel down the Mississippi to degrade coastal fisheries in the Gulf of Mexico. But other adverse effects are showing up within agricultural production systems -- for example, the rapidly developing resistance among pests is rendering our arsenal of herbicides and insecticides increasingly ineffective.". Agrochemicals and monoculture have been implicated in Colony Collapse Disorder, in which the individual members of bee colonies disappear. Agricultural production is highly dependent on bees to pollinate many varieties of fruits and vegetables.

Social

A study done for the US. Office of Technology Assessment conducted by the UC Davis Macrosocial Accounting Project concluded that industrial agriculture is associated with substantial deterioration of human living conditions in nearby rural communities.

Milpa

A typical modern Central American Milpa. The corn stalks have been bent and left to dry with cobs still on, for other crops, such as beans, to be planted. (Note: the banana plants in the background are not native, but are now a common part of modern Central American agriculture)

Milpa is a crop-growing system used throughout Mesoamerica. It has been most extensively described in the Yucatán peninsula area of Mexico. The word *milpa* is derived from the Nahuatl word phrase *mil-pa*, which translates into "maize field." Though different interpretations are given to it, it usually refers to a cropping field. Based on the ancient agricultural methods of Maya peoples and other Mesoamerican people, *milpa* agriculture produces maize, beans, and squash. The milpa cycle calls for 2 years of cultivation and eight years of letting the area lie fallow. Agronomists point out that the system is designed to create relatively large yields of food crops without the use of artificial pesticides or fertilizers, and they point out that while it is self-sustaining at current levels of consumption, there is a danger that at more intensive levels of cultivation the milpa system can become unsustainable.

The word is also used for a small field, especially in Mexico or Central America, that is cleared from the jungle, cropped for a few seasons, and then abandoned for a fresh clearing. In the states of Jalisco, Michoacán, and other areas of central Mexico, the term milpa simply means a single corn plant (milpas for plural). In El Salvador and Guatemala, it refers specifically to the corn crop or corn field as a whole.

Charles C. Mann described *milpa* agriculture as follows, in *1491: New Revelations of the Americas Before Columbus*:

"A milpa is a field, usually but not always recently cleared, in which farmers plant a dozen crops at once including maize, avocados, multiple varieties of squash and bean, melon, tomatoes, chilis, sweet potato, jícama, amaranth, and mucuna.... Milpa crops are nutritionally and environmentally complementary. Maize lacks the amino acids lysine and tryptophan, which the body needs to make proteins and niacin;.... Beans have both lysine and tryptophan.... Squashes, for their part, provide an array of vitamins; avocados, fats. The milpa, in the estimation of H. Garrison Wilkes, a maize researcher at the University of Massachusetts in Boston, "is one of the most successful human inventions ever created.""

The concept of *milpa* is a sociocultural construct rather than simply a system of agriculture. It involves complex interactions and relationships between farmers, as well as distinct personal relationships with both the crops and land. For example, it has been noted that "the making of *milpa* is the central, most sacred act, one which binds together the family, the community, the universe... [it] forms the core institution of Indian society in Mesoamerica and its religious and social importance often appear to exceed its nutritional and economic importance."

Milpitas, California derives its name from the Nahuatl term "milpa" followed by the Spanish feminine diminutive plural suffix "-itas".

Shifting Cultivation

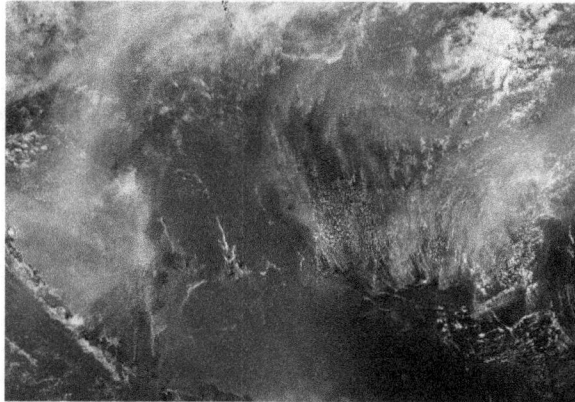

Slash-and-burn based shifting cultivation is a widespread historical practice in southeast Asia. Above is a satellite image of Sumatra and Borneo showing shift cultivation fires from October 2006.

Shifting cultivation is an agricultural system in which plots of land are cultivated temporarily, then abandoned and allowed to revert to their natural vegetation while the cultivator moves on to another plot. The period of cultivation is usually terminated when the soil shows signs of exhaustion or, more commonly, when the field is overrun by weeds. The length of time that a field is cultivated is usually shorter than the period over which the land is allowed to regenerate by lying fallow. This technique is often used in LEDCs (Less Economically Developed Countries) or LICs (Low Income Countries). In some areas, cultivators use a practice of slash-and-burn as one element of their farming cycle. Others employ land clearing without any burning, and some cultivators are purely migratory and do not use any cyclical method on a given plot. Sometimes no slashing at all is needed where regrowth is purely of grasses, an outcome not uncommon when soils are near exhaustion and need to lie fallow. In shifting agriculture, after two or three years of producing vegetable and grain crops on cleared land, the migrants abandon it for another plot. Trees, bushes and forests are cleared by slashing, and the remaining vegetation is burnt. The ashes add potash to the soil. Then the seeds are sown after the rains.

Advantages of Slash-and-burn Method

Slash-and-burn is a very sustainable technique. It differs a lot from commercial farming, because once the trees are burned, there is very fertile fine ash that deposits along the humus, meaning that by the time the other fields are burned, the soil has time to reassemble nutriments, in order to make cultural activity possible. Although slash-and-burn is a very useful technique, there are other ways of fertilizing soil. By planting beans, the soil will regenerate much faster, due to the production of nitrogen in their roots.

Political Ecology of Shifting Cultivation

Shifting cultivation is a form of agriculture or a cultivation system, in which, at any particular point in time, a minority of 'fields' are in cultivation and a majority are in various stages of natural regrowth. Over time, fields are cultivated for a relatively short time, and allowed to recover, or are fallowed, for a relatively long time. Eventually a previously cultivated field will be cleared of the natural vegetation and planted in crops again. Fields in established and stable shifting cultivation systems are cultivated and fallowed cyclically.This type of farming is called jhumming in India.

Fallow fields are not unproductive. During the fallow period, shifting cultivators use the successive vegetation species widely for timber for fencing and construction, firewood, thatching, ropes, clothing, tools, carrying devices and medicines. It is common for fruit and nut trees to be planted in fallow fields to the extent that parts of some fallows are in fact orchards. Soil-enhancing shrub or tree species may be planted or protected from slashing or burning in fallows. Many of these species have been shown to fix nitrogen. Fallows commonly contain plants that attract birds and animals and are important for hunting. But perhaps most importantly, tree fallows protect soil against physical erosion and draw nutrients to the surface from deep in the soil profile.

The relationship between the time the land is cultivated and the time it is fallowed are critical to the stability of shifting cultivation systems. These parameters determine whether or not the shifting cultivation system as a whole suffers a net loss of nutrients over time. A system in which there is a net loss of nutrients with each cycle will eventually lead to a degradation of resources unless actions are taken to arrest the losses. In some cases soil can be irreversibly exhausted (including erosion as well as nutrient loss) in less than a decade.

The longer a field is cropped, the greater the loss of soil organic matter, cation-exchange-capacity and in nitrogen and phosphorus, the greater the increase in acidity, the more likely soil porosity and infiltration capacity is reduced and the greater the loss of seeds of naturally occurring plant species from soil seed banks. In a stable shifting cultivation system, the fallow is long enough for the natural vegetation to recover to the state that it was in before it was cleared, and for the soil to recover to the condition it was in before cropping began. During fallow periods soil temperatures are lower, wind and water erosion is much reduced, nutrient cycling becomes closed again, nutrients are extracted from the subsoil, soil fauna decreases, acidity is reduced, soil structure, texture and moisture characteristics improve and seed banks are replenished.

The secondary forests created by shifting cultivation are commonly richer in plant and animal resources useful to humans than primary forests, even though they are much less bio-diverse. Shifting cultivators view the forest as an agricultural landscape of fields at various stages in a regular cycle. People unused to living in forests cannot see the fields for the trees. Rather they perceive an apparently chaotic landscape in which trees are cut and burned randomly and so they characterise shifting cultivation as ephemeral or 'pre-agricultural', as 'primitive' and as a stage to be progressed beyond. Shifting agriculture is none of these things. Stable shifting cultivation systems are highly variable, closely adapted to micro-environments and are carefully managed by farmers during both the cropping and fallow stages. Shifting cultivators may possess a highly developed knowledge and understanding of their local environments and of the crops and native plant species they exploit. Complex and highly adaptive land tenure systems sometimes exist under shifting cultivation. Introduced crops for food and as cash have been skillfully integrated into some shifting cultivation systems.

Shifting Cultivation in Europe

Shifting cultivation was still being practised as a viable and stable form of agriculture in many parts of Europe and east into Siberia at the end of the 19th century and in some places well into the 20th century. In the Ruhr in the late 1860s a forest-field rotation system known as *Reutberg-wirtschaft* was using a 16-year cycle of clearing, cropping and fallowing with trees to produce bark for tanneries, wood for charcoal and rye for flour (Darby 1956, 200). Swidden farming was practised in Siberia at least until the 1930s, using specially selected varieties of "swidden-rye" (Steensberg 1993, 98). In Eastern Europe and Northern Russia the main swidden crops were turnips, barley, flax, rye, wheat, oats, radishes and millet. Cropping periods were usually one year, but were extended to two or three years on very favourable soils. Fallow periods were between 20 and 40 years (Linnard 1970, 195). In Finland in 1949, Steensberg (1993, 111) observed the clearing and burning of a 60,000 square metres (15 acres) swidden 440 km north of Helsinki. Birch and pine trees had been cleared over a period of a year and the logs sold for cash. A fallow of alder (Alnus) was encouraged to improve soil conditions. After the burn, turnip was sown for sale and for cattle feed. Shifting cultivation was disappearing in this part of Finland because of a loss of agricultural labour to the industries of the towns. Steensberg (1993, 110-152) provides eye-witness descriptions of shifting cultivation being practised in Sweden in the 20th century, and in Estonia, Poland, the Caucasus, Serbia, Bosnia, Hungary, Switzerland, Austria and Germany in the 1930s to the 1950s.

That these agricultural practices survived from the Neolithic into the middle of the 20th century amidst the sweeping changes that occurred in Europe over that period, suggests they were adaptive and in themselves, were not massively destructive of the environments in which they were practised. This raises the question: if shifting cultivation did not lead to the disappearance of European forests, what did?

The earliest written accounts of forest destruction in Southern Europe begin around 1000 BC in the histories of Homer, Thucydides and Plato and in Strabo's Geography. Forests were exploited for ship building, and urban development, the manufacture of casks, pitch and charcoal, as well as being cleared for agriculture. The intensification of trade and as a result of warfare, increased the demand for ships which were manufactured completely from forest products. Although goat herding is singled out as an important cause of environmental degradation, a more important cause of forest destruction was the practice in some places of granting ownership rights to those who clear felled forests and brought the land into permanent cultivation. Evidence that circumstances other than agriculture were the major causes for forest destruction was the recovery of tree cover in many parts of the Roman empire from 400 BC to around 500 AD following the collapse of Roman economy and industry. Darby observes that by 400 AD "land that had once been tilled became derelict and overgrown" and quotes Lactantius who wrote that in many places "cultivated land became forest" (Darby 1956, 186). The other major cause of forest destruction in the Mediterrancan environment with its hot dry summers were wild fires that became more common following human interference in the forests.

In Central and Northern Europe the use of stone tools and fire in agriculture is well established in the palynological and archaeological record from the Neolithic. Here, just as in Southern Europe, the demands of more intensive agriculture and the invention of the plough, trading, mining and smelting, tanning, building and construction in the growing towns and constant warfare, including the demands of naval shipbuilding, were more important forces behind the destruction of the forests than was shifting cultivation.

By the Middle Ages in Europe, large areas of forest were being cleared and converted into arable land in association with the development of feudal tenurial practices. From the 16th to the 18th centuries, the demands of iron smelters for charcoal, increasing industrial developments and the discovery and expansion of colonial empires as well as incessant warfare that increased the demand for shipping to levels never previously reached, all combined to deforest Europe. With the loss of the forest, so shifting cultivation became restricted to the peripheral places of Europe, where permanent agriculture was uneconomic, transport costs constrained logging or terrain prevented the use of draught animals or tractors. It has disappeared from even these refuges since 1945, as agriculture has become increasingly capital intensive, rural areas have become depopulated and the remnant European forests themselves have been revalued economically and socially.

Simple Societies, Shifting Cultivation and Environmental Change

Shifting cultivation in Indonesia. A new crop is sprouting through the burnt soil.

A growing body of palynological evidence finds that simple human societies brought about extensive changes to their environments before the establishment of any sort of state, feudal or capitalist, and before the development of large scale mining, smelting or shipbuilding industries. In these societies agriculture was the driving force in the economy and shifting cultivation was the most common type of agriculture practiced. By examining the relationships between social and economic change and agricultural change in these societies, insights can be gained on contemporary social and economic change and global environment change, and the place of shifting cultivation in those relationship.

As early as 1930 questions about relationships between the rise and fall of the Mayan civilization of the Yucatán Peninsula and shifting cultivation were raised and continue to be debated today. Archaeological evidence suggests the development of Mayan society and economy began around 250 AD. A mere 700 years later it reached its apogee, by which time the population may have reached 2,000,000 people. There followed a precipitous decline that left the great cities and ceremonial centres vacant and overgrown with jungle vegetation. The causes of this decline are uncertain; but warfare and the exhaustion of agricultural land are commonly cited (Meggers 1954; Dumond 1961; Turner 1974). More recent work suggests the Maya may have, in suitable places, developed irrigation systems and more intensive agricultural practices (Humphries 1993).

Similar paths appear to have been followed by Polynesian settlers in New Zealand and the Pacific Islands, who within 500 years of their arrival around 1100 AD turned substantial areas from forest into scrub and fern and in the process caused the elimination of numerous species of birds and animals (Kirch and Hunt 1997). In the restricted environments of the Pacific islands, including Fiji and Hawaii, early extensive erosion and change of vegetation is presumed to have been caused by shifting cultivation on slopes. Soils washed from slopes were deposited in valley bottoms as a rich, swampy alluvium. These new environments were then exploited to develop intensive, irrigated fields. The change from shifting cultivation to intensive irrigated fields occurred in association with a rapid growth in population and the development of elaborate and highly stratified chiefdoms (Kirch 1984). In the larger, temperate latitude, islands of New Zealand the presumed course of events took a different path. There the stimulus for population growth was the hunting of large birds to extinction, during which time forests in drier areas were destroyed by burning, followed the development of intensive agriculture in favorable environments, based mainly on sweet potato (Ipomoea batatas) and a reliance on the gathering of two main wild plant species in less favorable environments. These changes, as in the smaller islands, were accompanied by population growth, the competition for the occupation of the best environments, complexity in social organization, and endemic warfare (Anderson 1997).

The record of humanly induced changes in environments is longer in New Guinea than in most places. Agricultural activities probably began 5,000 to 9,000 years ago. However, the most spectacular changes, in both societies and environments, are believed to have occurred in the central highlands of the island within the last 1,000 years, in association with the introduction of a crop new to New Guinea, the sweet potato (Golson 1982a; 1982b). One of the most striking signals of the relatively recent intensification of agriculture is the sudden increase in sedimentation rates in small lakes.

The root question posed by these and the numerous other examples that could be cited of simple societies that have intensified their agricultural systems in association with increases in population and social complexity is not whether or how shifting cultivation was responsible for the extensive changes to landscapes and environments. Rather it is why simple societies of shifting cultivators in the tropical forest of Yucatán, or the highlands of New Guinea, began to grow in numbers and to develop stratified and sometimes complex social hierarchies?

At first sight, the greatest stimulus to the intensification of a shifting cultivation system is a growth in population. If no other changes occur within the system, for each extra person to be fed from the system, a small extra amount of land must be cultivated. The total amount of land available is the land being presently cropped and all of the land in fallow. If the area occupied by the system is not expanded into previously unused land, then either the cropping period must be extended or the fallow period shortened.

At least two problems exist with the population growth hypothesis. First, population growth in most pre-industrial shifting cultivator societies has been shown to be very low over the long term. Second, no human societies are known where people work only to eat. People engage in social relations with each other and agricultural produce is used in the conduct of these relationships.

These relationships are the focus of two attempts to understand the nexus between human societies and their environments, one an explanation of a particular situation and the other a general exploration of the problem.

1. Feedback Loops

In a study of the Duna in the Southern Highlands of New Guinea, a group in the process of moving from shifting cultivation into permanent field agriculture post sweet potato, Modjeska (1982) argued for the development of two "self amplifying feed back loops" of ecological and social causation. The trigger to the changes was very slow population growth and the slow expansion of agriculture to meet the demands of this growth. This set in motion the first feedback loop, the "use-value" loop. As more forest was cleared there was a decline in wild food resources and protein produced from hunting, which was substituted for by an increase in domestic pig raising. An increase in domestic pigs required a further expansion in agriculture. The greater protein available from the larger number of pigs increased human fertility and survival rates and resulted in faster population growth.

The outcome of the operation of the two loops, one bringing about ecological change and the other social and economic change, is an expanding and intensifying agricultural system, the conversion of forest to grassland, a population growing at an increasing rate and expanding geographically and a society that is increasing in complexity and stratification.

2. Resources are Cultural Appraisals

The second attempt to explain the relationships between simple agricultural societies and their environments is that of Ellen (1982, 252-270). Ellen does not attempt to separate use-values from social production. He argues that almost all of the materials required by humans to live (with perhaps the exception of air) are obtained through social relations of production and that these relations proliferate and are modified in numerous ways. The values that humans attribute to items produced from the environment arise out of cultural arrangements and not from the objects themselves, a restatement of Carl Sauer's dictum that "resources are cultural appraisals". Humans frequently translate actual objects into culturally conceived forms, an example being the translation by the Duna of the pig into an item of compensation and redemption. As a result, two fundamental processes underlie the ecology of human social systems: First, the obtaining of materials from the environment and their alteration and circulation through social relations, and second, giving the material a value which will affect how important it is to obtain it, circulate it or alter it. Environmental pressures are thus mediated through social relations.

Transitions in ecological systems and in social systems do not proceed at the same rate. The rate of phylogenetic change is determined mainly by natural selection and partly by human interference and adaptation, such as for example, the domestication of a wild species. Humans however have the ability to learn and to communicate their knowledge to each other and across generations. If most social systems have the tendency to increase in complexity they will, sooner or later, come into conflict with, or into "contradiction" (Friedman 1979, 1982) with their environments. What happens around the point of "contradiction" will determine the extent of the environmental degradation that will occur. Of particular importance is the ability of the society to change, to invent or to innovate technologically and sociologically, in order to overcome the "contradiction" without incurring continuing environmental degradation, or social disintegration.

An economic study of what occurs at the points of conflict with specific reference to shifting cultivation is that of Esther Boserup (1965). Boserup argues that low intensity farming, extensive shifting cultivation for example, has lower labor costs than more intensive farming systems. This assertion remains controversial. She also argues that given a choice, a human group will always choose the technique which has the lowest absolute labor cost rather than the highest yield. But at the point of conflict, yields will have become unsatisfactory. Boserup argues, contra Malthus, that rather than population always overwhelming resources, that humans will invent a new agricultural technique or adopt an existing innovation that will boost yields and that is adapted to the new environmental conditions created by the degradation which has occurred already, even though they will pay for the increases in higher labor costs. Examples of such changes are the adoption of new higher yielding crops, the exchanging of a digging stick for a hoe, or a hoe for a plough, or the development of irrigation systems. The controversy over Boserup's proposal is in part over whether intensive systems are more costly in labor terms, and whether humans will bring about change in their agricultural systems before environmental degradation forces them to.

Shifting Cultivation in the Contemporary World and Global Environmental Change

Contemporary Shifting Cultivation Practice

Sumatra, Indonesia

Rio Xingu, Brazil

Santa Cruz, Bolivia

Kasempa , Zambia

The estimated rate of deforestation in Southeast Asia in 1990 was 34,000 km² per year (FAO 1990, quoted in Potter 1993). In Indonesia alone it was estimated 13,100 km² per year were being lost, 3,680 km² per year from Sumatra and 3,770 km² from Kalimantan, of which 1,440 km² were due to the fires of 1982 to 1983. Since those estimates were made huge fires have ravaged Indonesian forests during the 1997 to 1998 El Niño associated drought.

Interdisciplinary Project

Shifting cultivation used to be the backbone of smallholder agriculture throughout the tropics, but today it is abandoned in many places in favor of large scale cash crop production – e.g. for biofuels. The extent of these changes is not well documented because shifting cultivation land rarely appears on official maps and census data seldom identifies shifting cultivators. Moreover, the consequences of these changes for livelihoods (e.g. food security) are not well known. The aim of this project is to analyze the extent and consequences of change in shifting cultivation by combining meta-analyses of existing studies and census data with case studies in selected areas. This interdisciplinary project focuses on:

1. Trends in change in shifting cultivation landscapes and demography; and

2. Changes in livelihoods due to these changes.

The project will compile data for eight countries (Mexico, Brazil, Laos, Vietnam, Malaysia, Thailand, Zambia and Tanzania) and the outcome is expected to be relevant to planning and policy-making on land and forest management.

Shifting cultivation was assessed by the FAO to be one a causes of deforestation while logging was not. The apparent discrimination against shifting cultivators caused a confrontation between FAO and environmental groups, who saw the FAO supporting commercial logging interests against the rights of indigenous people (Potter 1993, 108). Other independent studies of the problem note that despite lack of government control over forests and the dominance of a political elite in the logging industry, the causes of deforestation are more complex. The loggers have provided paid employment to former subsistence farmers. One of the outcomes of cash incomes has been rapid population growth among indigenous groups of former shifting cultivators that has placed pressure on their traditional long fallow farming systems. Many farmers have taken advantage of the improved road access to urban areas by planting cash crops, such as rubber or pepper as noted above. Increased cash incomes often are spent on chain saws, which have enabled larger areas to be cleared for cultivation. Fallow periods have been reduced and cropping periods extended. Serious poverty elsewhere in the country has brought thousands of land hungry settlers into the cut over forests along the logging roads. The settlers practice what appears to be shifting cultivation but which is in fact a one-cycle slash and burn followed by continuous cropping, with no intention to long fallow. Clearing of trees and the permanent cultivation of fragile soils in a tropical environment with little attempt to replace lost nutrients may cause rapid degradation of the fragile soils.

The loss of forest in Indonesia, Thailand, and the Philippines during the 1990s was preceded by major ecosystem disruptions in Vietnam, Laos and Cambodia in the 1970s and 1980s caused by warfare. Forests were sprayed with defoliants, thousands of rural forest dwelling people uproots from their homes and moved and roads driven into previously isolated areas. The loss of the tropical forests of Southeast Asia is the particular outcome of the general possible outcomes described by Ellen when small local ecological and social systems become part of larger system. When the previous relatively stable ecological relationships are destabilized, degradation can occur rapidly. Similar descriptions of the loss of forest and destruction of fragile ecosystems could be provided from the Amazon Basin, by large scale state sponsored colonization forest land (Becker 1995, 61) or from the Central Africa where what endemic armed conflict is destabilizing rural settlement and farming communities on a massive scale.

Comparison with other Ecological Phenomena

In the tropical developing world, shifting cultivation in its many diverse forms, remains a pervasive practice. Shifting cultivation was one of the very first forms of agriculture practiced by humans and its survival into the modern world suggests that it is a flexible and highly adaptive means of production. However, it is also a grossly misunderstood practice. Many casual observers cannot see past the clearing and burning of standing forest and do not perceive often ecologically stable cycles of cropping and fallowing. Nevertheless, shifting cultivation systems are particularly susceptible to rapid increases in population and to economic and social change in the larger world around them. The blame for the destruction of forest resources is often laid on shifting cultivators. But the forces bringing about the rapid loss of tropical forests at the end of the 20th century are the same forces that led to the destruction of the forests of Europe, urbanization, industrialization, increased affluence, populational growth and geographical expansion and the application the latest technology to extract ever more resources from the environment in pursuit of wealth and political power by competing groups. However we must know that those who practice Agriculture are at the receiving end of the social stratum.

Studies of small, isolated and pre-capitalist groups and their relationships with their environments suggests that the roots of the contemporary problem lie deep in human behavioral patterns, for even in these simple societies, competition and conflict can be identified as the main force driving them into contradiction with their environments.

Alternative Practice in the pre-Columbian Amazon Basin

Slash-and-char, as opposed to slash-and-burn, may create self-perpetuating soil fertility that supports sedentary agriculture, but the society so sustained may still be overturned, as above .

Slash-and-burn

Slash-and-burn practices in Eno, Finland, 1893

This satellite photograph illustrates slash-and-burn forest clearing along the Rio Xingu (Xingu River) in the state of Mato Grosso, Brazil.

Slash-and-burn agriculture, or fire–fallow cultivation, is a farming method that involves the cutting and burning of plants in a forest or woodland to create a field called a *swidden*. (Preparing fields by deforestation is called *assarting*.) In subsistence agriculture, slash-and-burn typically uses little technology. It is often applied in shifting cultivation agriculture (such as in the Amazon rainforest) and in transhumance livestock herding.

Slash-and-burn is used by 200–500 million people worldwide. In 2004 it was estimated that in Brazil alone, 500,000 small farmers each cleared an average of one hectare (2.47105 acres) of forest per year. The technique is not scalable or sustainable for large human populations. Methods such as Inga alley farming have been proposed as alternatives to this environmental degradation.

History

Historically, slash-and-burn cultivation has been practiced throughout much of the world, in grasslands as well as woodlands.

During the Neolithic Revolution, which included agricultural advancements, groups of hunter-gatherers domesticated various plants and animals, permitting them to settle down and practice agriculture, which provides more nutrition per hectare than hunting and gathering. This happened in the river valleys of Egypt and Mesopotamia. Due to this decrease in food from hunting, as human populations increased, agriculture became more important. Some groups could easily plant their crops in open fields along river valleys, but others had forests blocking their farming land.

In this context, humans used slash-and-burn agriculture to clear more land to make it suitable for plants and animals. Thus, since Neolithic times, slash-and-burn techniques have been widely used for converting forests into crop fields and pasture. Fire was used before the Neolithic as well, and by hunter-gatherers up to present times. Clearings created by fire were made for many reasons, such as to draw game animals and to promote certain kinds of edible plants such as berries.

Slash-and-burn fields are typically used and owned by a family until the soil is exhausted. At this point the ownership rights are abandoned, the family clears a new field, and trees and shrubs are permitted to grow on the former field. After a few decades, another family or clan may then use the land and claim usufructuary rights. In such a system there is typically no market in farmland, so land is not bought or sold in the open market and land rights are traditional. In slash-and-burn agriculture, forests are typically cut months before a dry season. The "slash" is permitted to dry, and then burned in the following dry season. The resulting ash fertilizes the soil and the burned field is then planted at the beginning of the next rainy season with crops such as upland rice, maize, cassava, or other staples. Most of this work is typically done by hand, using such basic tools as machetes, axes, hoes, and makeshift shovels.

Large families or clans wandering in the lush woodlands long continued to be the most common form of life through human prehistory. Axes to fell trees and sickles for harvesting grain were the only tools people might bring with them. All other tools were made from materials they found at the site, such as fire stakes of birch, long rods (vanko), and harrows made of spruce tops. The extended family conquered the lush virgin forest, burned and cultivated their carefully selected swidden plots, sowed one or more crops, and then proceeded on to forests that had been noted in their wanderings. In the temperate zone the forest regenerated in the course of a lifetime. So swidden was repeated several times in the same area over the years. But in the tropics the forest floor gradually depleted. It was not only in the moors, as in Northern Europe, but also in the steppe, savannah, prairie, pampas and barren desert in tropical areas where shifting cultivation is the oldest type of farming (Clark 1952 91–107).

Historical Overview

Painting by Eero Järnefelt of forest-burning

Southern European Mediterranean climates have favored evergreen and deciduous forests. With slash-and-burn agriculture, this type of forest was less able to regenerate than those north of the Alps. Although in northern Europe one crop was usually harvested before grass was allowed to grow, in southern Europe it was more common to exhaust the soil by farming it for several years.

Classical authors mentioned large forests, with Homer writing about "wooded Samothrace," Zakynthos, Sicily, and other woodlands. These authors indicated that the Mediterranean area once had more forest; much had already been lost, and the remainder was primarily in the mountains.

Although parts of Europe aside from the north remained wooded, by the Roman Iron and early Viking Ages, forests were drastically reduced and settlements regularly moved. The reasons for this pattern of mobility, the transition to stable settlements from the late Viking period on, or the transition from shifting cultivation to stationary farming are unknown. From this period, plows are found in graves. Early agricultural peoples preferred good forests on hillsides with good drainage, and traces of cattle enclosures are evident there.

Greek explorer and merchant Pytheas of Marseilles made a voyage to Northern Europe around 330 BC, with part of his itinerary recorded by Polybios, Pliny, and Strabo. Pytheas visited Thule, a six-day voyage north of Britain: "The barbarians showed us the place where the sun does not go to sleep. It happened because there the night was very short—in some places two, in others three hours—so that the sun shortly after its fall soon went up again." He describes a fertile land, "rich in fruits that were ripe only until late in the year, and the people there used to prepare a drink of honey. And they threshed the grain in large houses because of the cloudy weather and frequent rain. In the spring they drove the cattle up into the mountain pastures and stayed there all summer."

In Italy, shifting cultivation was a thing of the past by the birth of Christ. Tacitus describes it as a strange cultivation method, practiced by the Germans. In 98 AD, he wrote about the Germans that their fields were proportional to the participating cultivators but their crops were shared according to status. Distribution was simple, because of wide availability; they changed fields annually, with much to spare because they were producing grain rather than other crops. According to the original text, "Agri pro numero cultorum ab universis in vices occupantur, quos mox inter se secundum dignationem partiuntur, facilitatem partiendi camporum spatia praestant. Arva per annos mutant, et superest ager; nec enim cum ubertate et amplitudine soli labore contendunt, ut pomaria conserant et prata separent et hortos rigent; sola terrae seges imperatur." This is the practice of shifting cultivation.

During the Migration Period in Europe, after the Roman Empire and before the Viking Age, the peoples of Central Europe moved to new forests after exhausting old parcels. Forests were quickly exhausted; the practice had ended in the Mediterranean, where forests were less resilient than the sturdier coniferous forests of Central Europe. Deforestation had been partially caused by burning to create pasture. Reduced timber delivery led to higher prices and more stone construction in the Roman Empire. Although forests gradually decreased in northern Europe, they have survived in the Nordic countries.

Tribes in pre-Roman Italy (including the Etruscans, Umbrians, Ligurians, Sabines, Latins, Campanians, Apulians, Saliscans, and Sabellians) apparently lived in temporary locations. They cultivated small patches of land, kept sheep and cattle, traded with foreign merchants, and occasionally fought. These Italic groups developed identities as settlers and warriors around 900 BC. They built forts in the mountains which are studied today, as are the ruins of a large Samnite temple and theater at Pietrabbondante.

Many Italic peoples saw benefits in allying with Rome. When the Romans built the *Via Amerina* in 241 BC, the Falisci settled in cities on the plains and aided the Romans in road construction; the Roman Senate gradually acquired representatives from Faliscan and Etruscan families, and the Italic tribes became settled farmers.

Classical writers described peoples who practiced shifting cultivation, which characterized the Migration Period in Europe. The exploitation of forests demanded displacement as areas were deforested. Julius Caesar wrote about the Suebi in *Commentarii de Bello Gallico* 4.1, "They have no private and secluded fields (*"privati ac separati agri apud eos nihil est"*) ... They cannot stay more than one year in a place for cultivation's sake" (*"neque longius anno remanere uno in loco colendi causa licet"*). The Suebi lived between the Rhine and the Elbe. About the Germani, Caesar wrote: "No one has a particular field or area for himself, for the magistrates and chiefs give year by year to the people and the clans, who have gathered together, as much land and in such places as seem good to them and then make them move on after a year" (*"Neque quisquam agri modum certum aut fines habet proprios, sed magistratus ac principes in annos singulos gentibus cognationibusque hominum, qui tum una coierunt, a quantum et quo loco visum est agri attribuunt atque anno post alio transire cogunt"*).

Strabo (63 BC—c. 20 AD) also writes about the Suebi in his *Geography* (VII, 1, 3): "Common to all the people in this area is that they can easily change residence because of their sordid way of life; they do not cultivate fields or collect property, but live in temporary huts. They get their nourishment from their livestock for the most part, and like nomads, pack all their goods in wagons and go on to wherever they want". Horace writes in 17 BC (*Carmen Saeculare*, 3, 24, 9ff.) about the people of Macedonia: "The proud Getae also live happily, growing free food and cereal for themselves on land they do not want to cultivate for more than a year" ("*Vivunt et rigidi Getae, / immetata quibus iugera liberas / fruges et Cererem ferunt, / nec cultura placet longior annua*").

Locations of Norwegian tribes described by Jordanes in his *Getica*

Jordanes, of Gothic descent, became a monk in Italy. In his mid-sixth-century AD *Getica* (*De origine actibusque Getarum*; *The Origin and Deeds of the Goths*), he described the large island of Scandza, on which the Goths originated. According to Jordanes, of the tribes living there, some are Adogit from within 40 days of the midnight sun. After the Adogit were the Screrefennae and Suehans, who also lived in the north. The Screrefennae did not raise crops, instead hunting and collecting bird eggs. The Suehans, a semi-nomadic tribe with good horses (comparable to the Thuringii), hunted furs to sell; grain could not be grown so far north. In about 550 AD, Procopius also described a primitive hunting people he called "Skrithifinoi": "Both men and women engaged incessantly just in hunting the rich forests and mountains, which gave them an endless supply of game and wild animals."

Slash-and-burn in Småland, Sweden (1904)

Adam of Bremen described Sweden from information he received from the Danish king Sven Estridson (also called Sweyn II of Denmark) in 1068: "It is very fruitful, the earth holds many crops and honey, it has greater livestock than all other countries, there are many useful rivers and forests; with regard to women they do not know moderation; they have for their households two, three, or more wives simultaneously; the rich and the rulers are innumerable … [with] live-stock grazing, as with the Arabs, far out in the wilderness". The use of fire in northeastern Sweden changed as agriculture evolved. Although the Sami people did not burn land (since burning killed the lichen required by their reindeer), later farmers frequently used slash-and-burn techniques. The 19th-century Swedish timber industry moved north, clearing the land of trees but leaving waste behind as a fire risk; during the 1870s, fires were frequent. There was a fire in Norrland in 1851, followed by fires in 1868 and 1878; two towns were lost in 1888.

Forest Finns

Huuhta cultivation spread: within the circle in 1500 AD, within the line in 1600, and to the dashed line in 1700.

One culture which flourished in pre-agricultural Europe survives: the Forest Finns in Scandinavia. Martin Tvengsberg, a descendant the Forest Finns, studied them in his capacity as curator of the Hedmark Museum in Norway. The Savo-Karelians had a sophisticated system for cultivating spruce forests. A runic poem about Finland's spruce forests reads, *"Gåivu on mehdien valgoinen valhe"* ("The birch is the forest's white lie").The best spruce forests reportedly contain birch trees, which grow only after a forest has burned once or twice.

Modern Western World

Slash-and-burn may be defined as the large-scale deforestation of forests for agricultural use. Ashes from the trees help farmers by providing nutrients for the soil.

In industrialized regions, including Europe and North America, the practice was abandoned with the introduction of market agriculture and land ownership. Slash-and-burn agriculture was initially practiced by European pioneers in North America such as Daniel Boone and his family, who cleared land in the Appalachian Mountains during the late 18th and early 19th centuries. However, land cleared by slash-and-burn farmers was eventually taken over by systems of land tenure focusing on long-term improvement and discouraging practices associated with slash-and-burn agriculture.

Northern European Heritage

Telkkämäki Heritage Farm and Nature Reserve in Kaavi, Finland

Some areas of the reserve are burned annually.

Telkkämäki Nature Reserve in Kaavi, Finland, is an open-air museum which still practices slash-and-burn agriculture. Farm visitors can see how people farmed when slash-and-burn agriculture became the norm in the Northern Savonian region of eastern Finland beginning in the 15th century. Areas of the reserve are burnt each year.

South Asia

Tribal groups in the northeastern Indian states of Arunachal Pradesh, Meghalaya, Mizoram and Nagaland and the Bangladeshi districts of Rangamati, Khagrachari, Bandarban and Sylhet refer to slash-and-burn agriculture as *jhum* or *jhoom* cultivation. The system involves clearing land, by fire or clear-felling, for economically-important crops such as upland rice, vegetables or fruits. After a few cycles, the land's fertility declines and a new area is chosen. *Jhum* cultivation is most often practiced on the slopes of thickly-forested hills. Cultivators cut the treetops to allow sunlight to reach the land, burning the trees and grasses for fresh soil. Although it is believed that this helps fertilize the land, it can leave it vulnerable to erosion. Holes are made for the seeds of crops such as sticky rice, maize, eggplant and cucumber are planted. After considering *jhum*'s effects, the government of Mizoram has introduced a policy to end the method in the state. Slash-and-burn is typically a type of subsistence agriculture, not focused on a need to sell crops globally; planting decisions are governed by the needs of the family (or clan) for the coming year.

Ecological Implications

Modern slash-and-burn practice

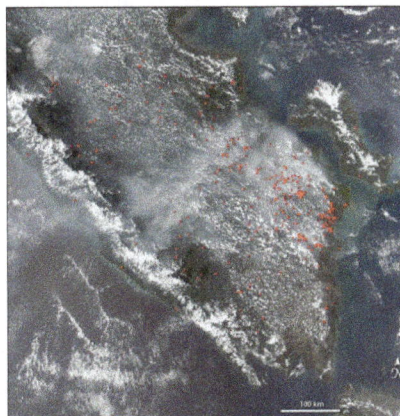

Sumatra, Indonesia

Chiang Mai, Thailand

Santa Cruz, Bolivia

Morondava, Madagascar

Although a solution for overpopulated tropical countries where subsistence agriculture may be the traditional method of sustaining many families, the consequences of slash-and-burn techniques for ecosystems are almost always destructive. This happens particularly as population densities increase, and as a result farming becomes more intensively practiced. This is because as demand for more land increases, the fallow period by necessity declines. The principal vulnerability is the nutrient-poor soil, pervasive in most tropical forests. When biomass is extracted even for one harvest of wood or charcoal, the residual soil value is heavily diminished for further growth of any type of vegetation.

Sometimes there are several cycles of slash-and-burn within a few years' time span. For example, in eastern Madagascar, the following scenario occurs commonly. The first wave might be cutting of all trees for wood use. A few years later, saplings are harvested to make charcoal, and within the next year the plot is burned to create a quick flush of nutrients for grass to feed the family zebu cattle. If adjacent plots are treated in a similar fashion, large-scale erosion will usually ensue, since there are no roots or temporary water storage in nearby canopies to arrest the surface runoff. Thus, any small remaining amounts of nutrients are washed away. The area is an example of desertification, and no further growth of any type may arise for generations.

The ecological ramifications of the above scenario are further magnified, because tropical forests are habitats for extremely biologically diverse ecosystems, typically containing large numbers of endemic and endangered species. Therefore, the role of slash-and-burn is significant in the current Holocene extinction.

Slash-and-char is an alternative that alleviates some of the negative ecological implications of traditional slash-and-burn techniques.

References

- G. Tyler Miller; Scott Spoolman (24 September 2008). Living in the Environment: Principles, Connections, and Solutions. Cengage Learning. pp. 279–. ISBN 978-0-495-55671-8. Retrieved 7 September 2010.

- Wilson, Gilbert (1917). Agriculture of the Hidatsa Indians: An Indian Interpretation. Gloucestershire: Dodo Press. p. 25. ISBN 978-1409942337.

- Mann, Charles (2005). 1491: New Revelations of the Americas Before Columbus. New York: Vintage Books. pp. 220–221. ISBN 978-1-4000-3205-1.

- Hemenway, Toby (2000). Gaia's Garden: A Guide to Home-Scale Permaculture. White River Junction, VT: Chelsea Green Publishing. p. 149. ISBN 1-890132-52-7.

- Johnson, Sue Ellen; Charles L. Mohler, (2009). Crop Rotation on Organic Farms: A Planning Manual, NRAES 177. Ithica, NY: National Resource, Agriculture, and Engineering Services (NRAES). ISBN 978-1-933395-21-0.

- Coleman, Eliot (1995), The New Organic Grower: A Master's Manual of Tools and Techniques for the Home and Market Gardener (2nd ed.), pp. 65, 108, ISBN 978-0930031756.

- Horne, Paul Anthony (2008). Integrated pest management for crops and pastures. CSIRO Publishing. p. 2. ISBN 978-0-643-09257-0.

- Vogt G (2007). Lockeretz W, ed. Chapter 1: The Origins of Organic Farming. Organic Farming: An International History. CABI Publishing. pp. 9–30. ISBN 9780851998336.

- Szykitka, Walter (2004). The Big Book of Self-Reliant Living: Advice and Information on Just About Everything You Need to Know to Live on Planet Earth. Globe-Pequot. p. 343. ISBN 978-1-59228-043-8.

- Pamela Ronald; Raoul Admachak (April 2008). "Tomorrow's Table: Organic Farming, Genetics and the Future of Food". Oxford University Press. ISBN 0195301757.

- Bridgewater, Samuel (2012). A Natural History of Belize: Inside the Maya Forest. London: Natural History Museum. pp. 154–155. ISBN 978-0-292-72671-0.

- Mann, Charles (2005). 1491: New Revelations of the Americas Before Columbus. New York: Vintage Books. pp. 197–198. ISBN 978-1-4000-3205-1.

- Waters, Tony (2007). The Persistence of Subsistence Agriculture. Lanham: Lexington Books. p. 3. ISBN 978-0-7391-0768-3. OCLC 70334845.

- Pyne, Stephen J. (1997). World Fire : The Culture of Fire on Earth (Pbk. ed.). Seattle: University of Washington Press. p. 85. ISBN 0295975938.

Agricultural Productivity: An Overview

The essential aspects of agricultural productivity included in this chapter are sustainable agriculture, green revolution, water resource management, nutrient management, ecological sanitation etc. Sustainable agriculture is a sustainable way of farming, and helps in studying the relation between organisms and their environment. The chapter discusses the aspects of agricultural productivity in a critical manner providing key analysis on the topic.

Agricultural Productivity

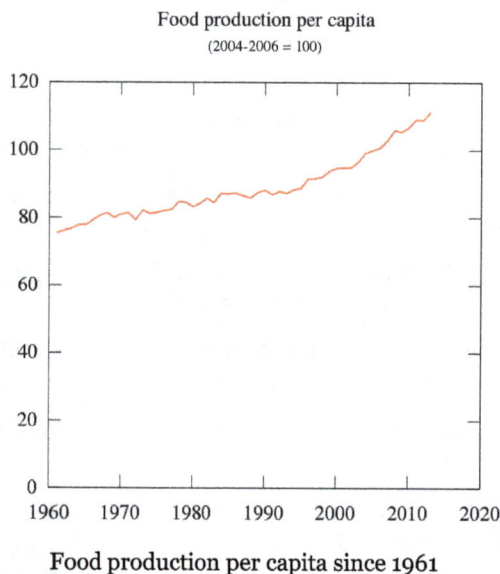

Food production per capita
(2004-2006 = 100)

Food production per capita since 1961

Agricultural productivity is measured as the ratio of agricultural outputs to agricultural inputs. While individual products are usually measured by weight, their varying densities make measuring overall agricultural output difficult. Therefore, output is usually measured as the market value of final output, which excludes intermediate products such as corn feed used in the meat industry. This output value may be compared to many different types of inputs such as labour and land (yield). These are called partial measures of productivity.

Agricultural productivity may also be measured by what is termed total factor productivity (TFP). This method of calculating agricultural productivity compares an index of agricultural inputs to an index of outputs. This measure of agricultural productivity was established to remedy the shortcomings of the partial measures of productivity; notably that it is often hard to identify the factors cause them to change. Changes in TFP are usually attributed to technological improvements.

Grain production facilities

Sources of agricultural productivity

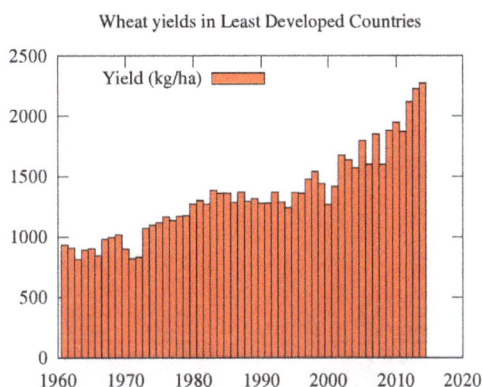

Wheat yields in least developed countries since 1961. The steep rise in crop yields in the U.S. began in the 1940s. The percentage of growth was fastest in the early rapid growth stage. In developing countries maize yields are still rapidly rising.

Some sources of agricultural productivity are:

- Mechanization

- High yield varieties, which were the basis of the Green revolution

- Fertilizers: Primary plant nutrients: nitrogen, phosphorus and potassium and secondary nutrients such as sulfur, zinc, copper, manganese, calcium, magnesium and molybdenum on deficient soil

- Liming of acid soils to raise pH and to provide calcium and magnesium

- Irrigation

- Herbicides

- Pesticides

- Increased plant density

- Animal feed made more digestible by processing

- Keeping animals indoors in cold weather

Importance of Agricultural Productivity

The productivity of a region's farms is important for many reasons. Aside from providing more food, increasing the productivity of farms affects the region's prospects for growth and competitiveness on the agricultural market, income distribution and savings, and labour migration. An increase in a region's agricultural productivity implies a more efficient distribution of scarce resources. As farmers adopt new techniques and differences, the more productive farmers benefit from an increase in their welfare while farmers who are not productive enough will exit the market to seek success elsewhere.

A cooperative dairy factory in Victoria.

As a region's farms become more productive, its comparative advantage in agricultural products increases, which means that it can produce these products at a lower opportunity cost than can other regions. Therefore, the region becomes more competitive on the world market, which means that it can attract more consumers since they are able to buy more of the products offered for the same amount of money.

Increases in agricultural productivity lead also to agricultural growth and can help to alleviate poverty in poor and developing countries, where agriculture often employs the greatest portion of the population. As farms become more productive, the wages earned by those who work in agriculture increase. At the same time, food prices decrease and food supplies become more stable. Labourers therefore have more money to spend on food as well as other products. This also leads to agricultural growth. People see that there is a greater opportunity to earn their living by farming and are attracted to agriculture either as owners of farms themselves or as labourers.

However, it is not only the people employed in agriculture who benefit from increases in agricultural productivity. Those employed in other sectors also enjoy lower food prices and a more stable food supply. Their wages may also increase.

A liquid manure spreader.

Agricultural productivity is becoming increasingly important as the world population continues to grow. India, one of the world's most populous countries, has taken steps in the past decades to increase its land productivity. Forty years ago, North India produced only wheat, but with the advent of the earlier maturing high-yielding wheats and rices, the wheat could be harvested in time to plant rice. This wheat/rice combination is now widely used throughout the Punjab, Haryana, and parts of Uttar Pradesh. The wheat yield of three tons and rice yield of two tons combine for five tons of grain per hectare, helping to feed India's 1.1 billion people.

Agricultural Productivity and Sustainable Development

Increase in agricultural productivity is often linked with questions about sustainability and sustainable development. Changes in agricultural practices necessarily bring changes in demands on resources. This means that as regions implement measures to increase the productivity of their farm land, they must also find ways to ensure that future generations will also have the resources they will need to live and thrive.

U.S. Agriculture Productivity

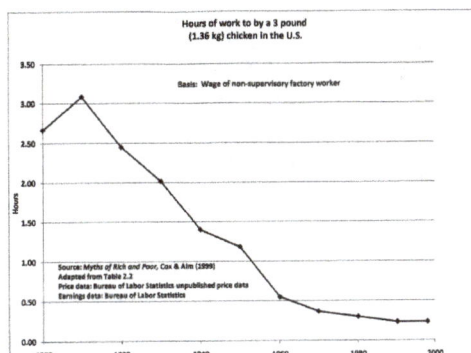

An hour's work in 1998 bought 11 times as much chicken as in 1900. Shown this way includes the increase in overall buying power plus lowered costs from the farm to the consumer.

Between 1950 and 2000, during the so-called "second agricultural revolution of modern times", U.S. agricultural productivity rose fast, especially due to the development of new technologies. For example, the average amount of milk produced per cow increased from 5,314 pounds to 18,201 pounds per year (+242%), the average yield of corn rose from 39 bushels to 153 bushels per acre (+292%), and each farmer in 2000 produced on average 12 times as much farm output per hour worked as a farmer did in 1950.

Productive Farms

Dairy cattle in Maryland

Some essential food products including bread, rice and pasta

For many farmers (especially in non-industrial countries) agricultural productivity may mean much more. A productive farm is one that provides most of the resources necessary for the farmer's family to live, such as food, fuel, fiber, healing plants, etc. It is a farm which ensures food security as well as a way to sustain the well-being of a community. This implies that a productive farm is also one which is able to ensure proper management of natural resources, such as biodiversity, soil, water, etc. For most farmers, a productive farm would also produce more goods than required for the community in order to allow trade.

Diversity in agricultural production is one key to productivity, as it enables risk management and preserves potentials for adaptation and change. Monoculture is an example of such a nondiverse production system. In a monocultural system a farmer may produce only crops, but no livestock, or only livestock and no crop.

The benefits of raising livestock, among others, are that it provides multiple goods, such as food, wool, hides, and transportation. It also has an important value in term of social relationships (such as gifts in weddings). In case of famine, when crops are not sufficient to ensure food safety, livestock can be used as food. Livestock may also provide manure, which can be used to fertilize cultivated soils, which increases soil productivity. On the other hand, in an agricultural system based only on raising livestock, food has to be bought from other farmers, and wastes produced cannot be easily disposed of. Production has many functions, and diversity is the foundation of such production. To ignore the complex functions provided by a farm is thought by many to turn agricultural production into a commodity.

Sustainable Agriculture

Sustainable agriculture is farming in sustainable ways based on an understanding of ecosystem services, the study of relationships between organisms and their environment. It has been defined as "an integrated system of plant and animal production practices having a site-specific application that will last over the long term", for example:

- Satisfy human food and fiber needs

- Enhance environmental quality and the natural resource base upon which the agricultural economy depends

- Make the most efficient use of non-renewable resources and on-farm resources and integrate, where appropriate, natural biological cycles and controls

- Sustain the economic viability of farm operations

- Enhance the quality of life for farmers and society as a whole

History of the Term

The phrase was reportedly coined by the Australian agricultural scientist Gordon McClymont. Wes Jackson is credited with the first publication of the expression in his 1980 book *New Roots for Agriculture*. The term became popularly used in the late 1980s.

Farming and Natural Resources

Sustainable agriculture can be understood as an ecosystem approach to agriculture. Practices that can cause long-term damage to soil include excessive tilling of the soil (leading to erosion) and irrigation without adequate drainage (leading to salinization). Long-term experiments have provided some of the best data on how various practices affect soil properties essential to sustainability. In the United States a federal agency, USDA-Natural Resources Conservation Service, specializes in providing technical and financial assistance for those interested in pursuing natural resource conservation and production agriculture as compatible goals.

Traditional farming methods had zero carbon footprint.

The most important factors for an individual site are sun, air, soil, nutrients, and water. Of the five, water and soil quality and quantity are most amenable to human intervention through time and labor.

Although air and sunlight are available everywhere on Earth, crops also depend on soil nutrients and the availability of water. When farmers grow and harvest crops, they remove some of these nutrients from the soil. Without replenishment, land suffers from nutrient depletion and becomes either unusable or suffers from reduced yields. Sustainable agriculture depends on replenishing the soil while minimizing the use or need of non-renewable resources, such as natural gas (used in converting atmospheric nitrogen into synthetic fertilizer), or mineral ores (e.g., phosphate). Possible sources of nitrogen that would, in principle, be available indefinitely, include:

1. recycling crop waste and livestock or treated human manure

2. growing legume crops and forages such as peanuts or alfalfa that form symbioses with nitrogen-fixing bacteria called rhizobia

3. industrial production of nitrogen by the Haber process uses hydrogen, which is currently derived from natural gas (but this hydrogen could instead be made by electrolysis of water using electricity (perhaps from solar cells or windmills)) or

4. genetically engineering (non-legume) crops to form nitrogen-fixing symbioses or fix nitrogen without microbial symbionts.

The last option was proposed in the 1970s, but is only recently becoming feasible. Sustainable options for replacing other nutrient inputs (phosphorus, potassium, etc.) are more limited.

More realistic, and often overlooked, options include long-term crop rotations, returning to natural cycles that annually flood cultivated lands (returning lost nutrients indefinitely) such as the flooding of the Nile, the long-term use of biochar, and use of crop and livestock landraces that are adapted to less than ideal conditions such as pests, drought, or lack of nutrients.

Crops that require high levels of soil nutrients can be cultivated in a more sustainable manner if certain fertilizer management practices are adhered to.

Nationwide food producers require vast amounts of land and soil to produce food at an accelerated rate. This diminishes the nutrients in the soil and decimates the idea of sustainable agriculture, which is best built through local, regional agricultural methods.

Water

In some areas sufficient rainfall is available for crop growth, but many other areas require irrigation. For irrigation systems to be sustainable, they require proper management (to avoid salinization) and must not use more water from their source than is naturally replenishable. Otherwise, the water source effectively becomes a non-renewable resource. Improvements in water well drilling technology and submersible pumps, combined with the development of drip irrigation and low-pressure pivots, have made it possible to regularly achieve high crop yields in areas where reliance on rainfall alone had previously made successful agriculture unpredictable. However, this progress has come at a price. In many areas, such as the Ogallala Aquifer, the water is being used faster than it can be replenished.

Several steps must be taken to develop drought-resistant farming systems even in "normal" years with average rainfall. These measures include both policy and management actions:

1. improving water conservation and storage measures,

2. providing incentives for selection of drought-tolerant crop species,

3. using reduced-volume irrigation systems,

4. managing crops to reduce water loss, and

5. not planting crops at all.

Indicators for sustainable water resource development are:

- Internal renewable water resources. This is the average annual flow of rivers and groundwater generated from endogenous precipitation, after ensuring that there is no double counting. It represents the maximum amount of water resource produced within the boundaries of a country. This value, which is expressed as an average on a yearly basis, is invariant in time (except in the case of proved climate change). The indicator can be expressed in three different units: in absolute terms (km^3/yr), in mm/yr (it is a measure of the humidity of the country), and as a function of population (m^3/person per year).

- Global renewable water resources. This is the sum of internal renewable water resources and incoming flow originating outside the country. Unlike internal resources, this value can vary with time if upstream development reduces water availability at the border. Treaties ensuring a specific flow to be reserved from upstream to downstream countries may be taken into account in the computation of global water resources in both countries.

- Dependency ratio. This is the proportion of the global renewable water resources originating outside the country, expressed in percentage. It is an expression of the level to which the water resources of a country depend on neighbouring countries.

- Water withdrawal. In view of the limitations described above, only gross water withdrawal can be computed systematically on a country basis as a measure of water use. Absolute or per-person value of yearly water withdrawal gives a measure of the importance of water in the country's economy. When expressed in percentage of water resources, it shows the degree of pressure on water resources. A rough estimate shows that if water withdrawal exceeds a quarter of global renewable water resources of a country, water can be considered a limiting factor to development and, reciprocally, the pressure on water resources can affect all sectors, from agriculture to environment and fisheries.

Soil

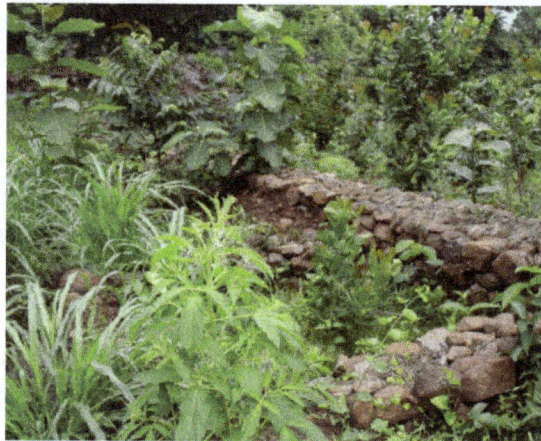

Walls built to avoid water run-off

Soil erosion is fast becoming one of the world's severe problems. It is estimated that "more than a thousand million tonnes of southern Africa's soil are eroded every year. Experts predict that crop yields will be halved within thirty to fifty years if erosion continues at present rates." Soil erosion is not unique to Africa but is occurring worldwide. The phenomenon is being called *peak soil* as present large-scale factory farming techniques are jeopardizing humanity's ability to grow food in the present and in the future. Without efforts to improve soil management practices, the availability of arable soil will become increasingly problematic.

Some soil management techniques

- No-till farming
- Keyline design
- Growing windbreaks to hold the soil
- Incorporating organic matter back into fields
- Stop using chemical fertilizers (which contain salt)
- Protecting soil from water run-off (soil erosion)

Phosphate

Phosphate is a primary component in the chemical fertilizer which is applied in modern agricultural production. However, scientists estimate that rock phosphate reserves will be depleted in 50–100 years and that peak phosphorus will occur in about 2030. The phenomenon of peak phosphorus is expected to increase food prices as fertilizer costs increase as rock phosphate reserves become more difficult to extract. In the long term, phosphate will therefore have to be recovered and recycled from human and animal waste in order to maintain food production.

Land

As the global population increases and demand for food increases, there is pressure on land resources. Land can also be considered a finite resource on Earth. Expansion of agricultural land decreases biodiversity and contributes to deforestation. The Food and Agriculture Organisation of the United Nations estimates that in coming decades, cropland will continue to be lost to industrial and urban development, along with reclamation of wetlands, and conversion of forest to cultivation, resulting in the loss of biodiversity and increased soil erosion.

Energy for Agriculture

Energy is used all the way down the food chain from farm to fork. In industrial agriculture, energy is used in on-farm mechanisation, food processing, storage, and transportation processes. It has therefore been found that energy prices are closely linked to food prices. Oil is also used as an input in agricultural chemicals. Higher prices of non-renewable energy resources are projected by the International Energy Agency. Increased energy prices as a result of fossil fuel resources being depleted may therefore decrease global food security unless action is taken to 'decouple' fossil fuel energy from food production, with a move towards 'energy-smart' agricultural systems. The use of solar powered irrigation in Pakistan has come to be recognized as a leading example of energy use in creating a closed system for water irrigation in agricultural activity.

Economics

Socioeconomic aspects of sustainability are also partly understood. Regarding less concentrated farming, the best known analysis is Netting's study on smallholder systems through history. The Oxford Sustainable Group defines sustainability in this context in a much broader form, considering effect on all stakeholders in a 360 degree approach

Given the finite supply of natural resources at any specific cost and location, agriculture that is inefficient or damaging to needed resources may eventually exhaust the available resources or the ability to afford and acquire them. It may also generate negative externality, such as pollution as well as financial and production costs. There are several studies incooperating these negative externalities in an economic analysis concerning ecosystem services, biodiversity, land degradation and sustainable land management. These include the The Economics of Ecosystems and Biodiversity (TEEB) study led by Pavan Sukhdev and the Economics of Land Degradation Initiative which seeks to establish an economic cost benefit analysis on the practice of sustainable land management and sustainable agriculture.

The way that crops are sold must be accounted for in the sustainability equation. Food sold locally does not require additional energy for transportation (including consumers). Food sold at a remote location, whether at a farmers' market or the supermarket, incurs a different set of energy cost for materials, labour, and transport.

Pursuing sustainable agriculture results in many localized benefits. Having the opportunities to sell products directly to consumers, rather than at wholesale or commodity prices, allows farmers to bring in optimal profit.

Methods

Polyculture practices in Andhra Pradesh

What grows where and how it is grown are a matter of choice. Two of the many possible practices of sustainable agriculture are crop rotation and soil amendment, both designed to ensure that crops being cultivated can obtain the necessary nutrients for healthy growth. Soil amendments would include using locally available compost from community recycling centers. These community recycling centers help produce the compost needed by the local organic farms.

Using community recycling from yard and kitchen waste utilizes a local area's commonly available resources. These resources in the past were thrown away into large waste disposal sites, are now used to produce low cost organic compost for organic farming. Other practices includes growing a diverse number of perennial crops in a single field, each of which would grow in separate season so as not to compete with each other for natural resources. This system would result in increased resistance to diseases and decreased effects of erosion and loss of nutrients in soil. Nitrogen fixation from legumes, for example, used in conjunction with plants that rely on nitrate from soil for growth, helps to allow the land to be reused annually. Legumes will grow for a season and replenish the soil with ammonium and nitrate, and the next season other plants can be seeded and grown in the field in preparation for harvest.

Rotational grazing practices in use with paddocks

Monoculture, a method of growing only one crop at a time in a given field, is a very widespread practice, but there are questions about its sustainability, especially if the same crop is grown every year. Today it is realized to get around this problem local cities and farms can work together to produce the needed compost for the farmers around them. This combined with growing a mixture of crops (polyculture) sometimes reduces disease or pest problems but polyculture has rarely, if ever, been compared to the more widespread practice of growing different crops in successive years (crop rotation) with the same overall crop diversity. Cropping systems that include a variety of crops (polyculture and/or rotation) may also replenish nitrogen (if legumes are included) and may also use resources such as sunlight, water, or nutrients more efficiently (Field Crops Res. 34:239).

Replacing a natural ecosystem with a few specifically chosen plant varieties reduces the genetic diversity found in wildlife and makes the organisms susceptible to widespread disease. The Great Irish Famine (1845–1849) is a well-known example of the dangers of monoculture. In practice, there is no single approach to sustainable agriculture, as the precise goals and methods must be adapted to each individual case. There may be some techniques of farming that are inherently in conflict with the concept of sustainability, but there is widespread misunderstanding on effects of some practices. Today the growth of local farmers' markets offer small farms the ability to sell the products that they have grown back to the cities that they got the recycled compost from. By using local recycling this will help move people away from the slash-and-burn techniques that are the characteristic feature of shifting cultivators are often cited as inherently destructive, yet slash-and-burn cultivation has been practiced in the Amazon for at least 6000 years; serious deforestation did not begin until the 1970s, largely as the result of Brazilian government programs and policies. To note that it may not have been slash-and-burn so much as slash-and-char, which with the addition of organic matter produces terra preta, one of the richest soils on Earth and the only one that regenerates itself.

There are also many ways to practice sustainable animal husbandry. Some of the key tools to grazing management include fencing off the grazing area into smaller areas called paddocks, lowering stock density, and moving the stock between paddocks frequently.

Sustainable Intensification

In light of concerns about food security, human population growth and dwindling land suitable for agriculture, sustainable intensive farming practises are needed to maintain high crop yields, while maintaining soil health and ecosystem services. The capacity for ecosystem services to be strong enough to allow a reduction in use of synthetic, non renewable inputs whilst maintaining or even boosting yields has been the subject of much debate. Recent work in the globally important irrigated rice production system of east Asia has suggested that - in relation to pest management at least - promoting the ecosystem service of biological control using nectar plants can reduce the need for insecticides by 70% whilst delivering a 5% yield advantage compared with standard practice.

Soil Treatment

Soil steaming can be used as an ecological alternative to chemicals for soil sterilization. Different methods are available to induce steam into the soil in order to kill pests and increase soil health.

Solarizing is based on the same principle, used to increase the temperature of the soil to kill pathogens and pests.

Sheet steaming with a MSD/moeschle steam boiler (left side)

Certain crops act as natural biofumigants, releasing pest suppressing compounds. Mustard, radishes, and other plants in the brassica family are best known for this effect. There exist varieties of mustard shown to be almost as effective as synthetic fumigants at a similar or lesser cost.

Off-farm Impacts

A farm that is able to "produce perpetually", yet has negative effects on environmental quality elsewhere is not sustainable agriculture. An example of a case in which a global view may be warranted is over-application of synthetic fertilizer or animal manures, which can improve productivity of a farm but can pollute nearby rivers and coastal waters (eutrophication). The other extreme can also be undesirable, as the problem of low crop yields due to exhaustion of nutrients in the soil has been related to rainforest destruction, as in the case of slash and burn farming for livestock feed.In Asia, specific land for sustainable farming is about 12.5 acres which includes land for animal fodder, cereals productions lands for some cash crops and even recycling of related food crops.In some cases even a small unit of aquaculture is also included in this number (AARI-1996)

Sustainability affects overall production, which must increase to meet the increasing food and fiber requirements as the world's human population expands to a projected 9.3 billion people by 2050. Increased production may come from creating new farmland, which may ameliorate carbon dioxide emissions if done through reclamation of desert as in Israel and Palestine, or may worsen emissions if done through slash and burn farming, as in Brazil.

International Policy

Sustainable agriculture has become a topic of interest in the international policy arena, especially with regards to its potential to reduce the risks associated with a changing climate and growing human population.

The Commission on Sustainable Agriculture and Climate Change, as part of its recommendations for policy makers on achieving food security in the face of climate change, urged that sustainable agriculture must be integrated into national and international policy. The Commission stressed that increasing weather variability and climate shocks will negatively affect agricultural yields, necessitating early action to drive change in agricultural production systems towards increasing resilience. It also called for dramatically increased investments in sustainable agriculture in the next decade, including in national research and development budgets, land rehabilitation, economic incentives, and infrastructure improvement.

Urban Planning

There has been considerable debate about which form of human residential habitat may be a better social form for sustainable agriculture.

Many environmentalists advocate urban developments with high population density as a way of preserving agricultural land and maximizing energy efficiency. However, others have theorized that sustainable ecocities, or ecovillages which combine habitation and farming with close proximity between producers and consumers, may provide greater sustainability.

The use of available city space (e.g., rooftop gardens, community gardens, garden sharing, and other forms of urban agriculture) for cooperative food production is another way to achieve greater sustainability.

One of the latest ideas in achieving sustainable agriculture involves shifting the production of food plants from major factory farming operations to large, urban, technical facilities called vertical farms. The advantages of vertical farming include year-round production, isolation from pests and diseases, controllable resource recycling, and on-site production that reduces transportation costs. While a vertical farm has yet to become a reality, the idea is gaining momentum among those who believe that current sustainable farming methods will be insufficient to provide for a growing global population.

Criticism

Efforts toward more sustainable agriculture are supported in the sustainability community, however, these are often viewed only as incremental steps and not as an end. Some foresee a true sustainable steady state economy that may be very different from today's: greatly reduced energy usage, minimal ecological footprint, fewer consumer packaged goods, local purchasing with short food supply chains, little processed foods, more home and community gardens, etc. Agriculture would be very different in this type of sustainable economy.

Green Revolution

The Green Revolution refers to a set of research and development of technology transfer initiatives occurring between the 1930s and the late 1960s (with prequels in the work of the agrarian geneticist Nazareno Strampelli in the 1920s and 1930s), that increased agricultural production worldwide, particularly in the developing world, beginning most markedly in the late 1960s. The initiatives resulted in the adoption of new technologies, including:

...new, high-yielding varieties (HYVs) of cereals, especially dwarf wheats and rices, in association with chemical fertilizers and agro-chemicals, and with controlled water-supply (usually involving irrigation) and new methods of cultivation, including mechanization. All of these together were seen as a 'package of practices' to supersede 'traditional' technology and to be adopted as a whole.

The initiatives, led by Norman Borlaug, the "Father of the Green Revolution," who received the Nobel Peace Prize in 1970, credited with saving over a billion people from starvation, involved the development of high-yielding varieties of cereal grains, expansion of irrigation infrastructure, modernization of management techniques, distribution of hybridized seeds, synthetic fertilizers, and pesticides to farmers.

The term "Green Revolution" was first used in 1968 by former US Agency for International Development (USAID) director William Gaud, who noted the spread of the new technologies: "These and other developments in the field of agriculture contain the makings of a new revolution. It is not a violent Red Revolution like that of the Soviets, nor is it a White Revolution like that of the Shah of Iran. I call it the Green Revolution."

History

Green Revolution in Mexico

It has been argued that "during the twentieth century two 'revolutions' transformed rural Mexico: the Mexican Revolution (1910-1920) and the Green Revolution (1950-1970). With the support of the Mexican government, the U.S. government, the United Nations, the Food and Agriculture Organization (FAO), and the Rockefeller Foundation, Mexico made a concerted effort to transform agricultural productivity, particularly with irrigated rather than dry-land cultivation in its northwest, to solve its problem of lack of food self-sufficiency. In the center and south of Mexico, where large-scale production faced challenges, agricultural production languished. Increased production meant food self-sufficiency in Mexico to feed its growing and urbanizing population, with the number of calories consumed per Mexican increasing. Technology was seen as a valuable way to feed the poor, and would relieve some pressure of the land redistribution process.

Mexico was not merely the recipient of Green Revolution knowledge and technology, but was an active participant with financial support from the government for agriculture as well as Mexican agronomists (*agrónomos*). Although the Mexican Revolution had broken the back of the hacienda system and land reform in Mexico had by 1940 distributed a large expanse of land in central and southern Mexico, agricultural productivity had fallen. During the administration of Manuel Avila Camacho (1940–46), the government put resources into developing new breeds of plants and partnered with the Rockefeller Foundation. In 1943, the Mexican government founded the International Maize and Wheat Improvement Center (CIMMYT), which became a base for international agricultural research.

Agriculture in Mexico had been a sociopolitical issue, a key factor in some regions' participation in the Mexican Revolution. It was also a technical issue, which the development of a cohort trained agronomists, who were to advise peasants how to increase productivity. In the post-World War II era, the government sought development in agriculture that bettered technological aspects of agriculture in regions that were not dominated by small-scale peasant cultivators. This drive for transforming agriculture would have the benefit of keeping Mexico self-sufficient in food and in the political sphere with the Cold War, potentially stem unrest and the appeal of Communism. Technical aid can be seen as also serving political ends in the international sphere. In Mexico, it also served political ends, separating peasant agriculture based on the ejido and considered one of the victories of the Mexican Revolution, from agribusiness that requires large-scale land ownership, irrigation, specialized seeds, fertilizers, and pesticides, machinery, and a low-wage paid labor force.

The government created the Mexican Agricultural Program (MAP) to be the lead organization in raising productivity. One of their successes was wheat production, with varieties the agency's scientists helped create dominating wheat production as early as 1951 (70%), 1965 (80%), and 1968 (90%). Mexico became the showcase for extending the Green Revolution to other areas of Latin America and beyond, into Africa and Asia. New breeds of maize, beans, along with wheat produced bumper crops with proper inputs (such as fertilizer and pesticides) and careful cultivation. Many Mexican farmers who had been dubious about the scientists or hostile to them (often a mutual relationship of discord) came to see the scientific approach to agriculture worth adopting.

Green Revolution in Rice: IR8 and the Philippines

In 1960, the Government of the Republic of the Philippines with the Ford Foundation and the Rockefeller Foundation established IRRI (International Rice Research Institute). A rice crossing between Dee-Geo-woo-gen and Peta was done at IRRI in 1962. In 1966, one of the breeding lines became a new cultivar, IR8. IR8 required the use of fertilizers and pesticides, but produced substantially higher yields than the traditional cultivars. Annual rice production in the Philippines increased from 3.7 to 7.7 million tons in two decades. The switch to IR8 rice made the Philippines a rice exporter for the first time in the 20th century.

Green Revolution's Start in India

In 1961, India was on the brink of mass famine. Norman Borlaug was invited to India by the adviser to the Indian minister of agriculture C. Subramaniam. Despite bureaucratic hurdles imposed by India's grain monopolies, the Ford Foundation and Indian government collaborated to import wheat seed from the International Maize and Wheat Improvement Center (CIMMYT). Punjab was selected by the Indian government to be the first site to try the new crops because of its reliable water supply and a history of agricultural success. India began its own Green Revolution program of plant breeding, irrigation development, and financing of agrochemicals.

India soon adopted IR8 – a semi-dwarf rice variety developed by the International Rice Research Institute (IRRI) that could produce more grains of rice per plant when grown with certain fertilizers and irrigation. In 1968, Indian agronomist S.K. De Datta published his findings that IR8 rice yielded about 5 tons per hectare with no fertilizer, and almost 10 tons per hectare under optimal conditions. This was 10 times the yield of traditional rice. IR8 was a success throughout Asia, and dubbed the "Miracle Rice." IR8 was also developed into Semi-dwarf IR36.

In the 1960s, rice yields in India were about two tons per hectare; by the mid-1990s, they had risen to six tons per hectare. In the 1970s, rice cost about $550 a ton; in 2001, it cost under $200 a ton. India became one of the world's most successful rice producers, and is now a major rice exporter, shipping nearly 4.5 million tons in 2006.

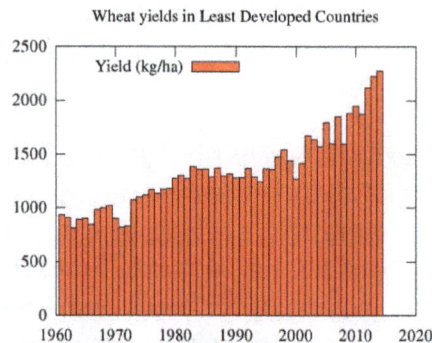

Wheat yields in least developed countries since 1961, in kilograms per hectare.

Consultative Group on International Agricultural Research - CGIAR

In 1970, foundation officials proposed a worldwide network of agricultural research centers under a permanent secretariat. This was further supported and developed by the World Bank; on 19 May 1971, the Consultative Group on International Agricultural Research(CGIAR) was established. co-sponsored by the FAO, IFAD and UNDP. CGIAR has added many research centers throughout the world.

CGIAR has responded, at least in part, to criticisms of Green Revolution methodologies. This began in the 1980s, and mainly was a result of pressure from donor organizations. Methods like Agroecosystem Analysis and Farming System Research have been adopted to gain a more holistic view of agriculture.

Brazil's Agricultural Revolution

Brazil's vast inland cerrado region was regarded as unfit for farming before the 1960s because the soil was too acidic and poor in nutrients, according to Norman Borlaug. However, from the 1960s, vast quantities of lime (pulverised chalk or limestone) were poured on the soil to reduce acidity. The effort went on for decades; by late 1990s, between 14 million and 16 million tonnes of lime were being spread on Brazilian fields each year. The quantity rose to 25 million tonnes in 2003 and 2004, equalling around five tonnes of lime per hectare. As a result, Brazil has become the world's second biggest soybean exporter. Soybeans are also widely used in animal feed, and the large volume of soy produced in Brazil has contributed to Brazil's rise to become the biggest exporter of beef and poultry in the world. Several parallels can also be found in Argentina's boom in soybean production as well.

Problems in Africa

There have been numerous attempts to introduce the successful concepts from the Mexican and Indian projects into Africa. These programs have generally been less successful. Reasons cited include widespread corruption, insecurity, a lack of infrastructure, and a general lack of will on the part of the governments. Yet environmental factors, such as the availability of water for irrigation, the high diversity in slope and soil types in one given area are also reasons why the Green Revolution is not so successful in Africa.

A recent program in western Africa is attempting to introduce a new high-yielding 'family' of rice varieties known as "New Rice for Africa" (NERICA). NERICA varieties yield about 30% more rice under normal conditions, and can double yields with small amounts of fertilizer and very basic irrigation. However, the program has been beset by problems getting the rice into the hands of farmers, and to date the only success has been in Guinea, where it currently accounts for 16% of rice cultivation.

After a famine in 2001 and years of chronic hunger and poverty, in 2005 the small African country of Malawi launched the "Agricultural Input Subsidy Program" by which vouchers are given to smallholder farmers to buy subsidized nitrogen fertilizer and maize seeds. Within its first year, the program was reported to have had extreme success, producing the largest maize harvest of the country's history, enough to feed the country with tons of maize left over. The program has advanced yearly ever since. Various sources claim that the program has been an unusual success, hailing it as a "miracle."

Agricultural Production and Food Security

Technologies

New varieties of wheat and other grains were instrumental to the green revolution.

The Green Revolution spread technologies that already existed, but had not been widely implemented outside industrialized nations. These technologies included modern irrigation projects, pesticides, synthetic nitrogen fertilizer and improved crop varieties developed through the conventional, science-based methods available at the time.

The novel technological development of the Green Revolution was the production of novel wheat cultivars. Agronomists bred cultivars of maize, wheat, and rice that are generally referred to as HYVs or "high-yielding varieties." HYVs have higher nitrogen-absorbing potential than other varieties. Since cereals that absorbed extra nitrogen would typically lodge, or fall over before harvest, semi-dwarfing genes were bred into their genomes. A Japanese dwarf wheat cultivar (Norin 10 wheat), which was sent to Washington, D.C. by Cecil Salmon, was instrumental in developing Green Revolution wheat cultivars. IR8, the first widely implemented HYV rice to be developed by IRRI, was created through a cross between an Indonesian variety named "Peta" and a Chinese variety named "Dee-geo-woo-gen."

With advances in molecular genetics, the mutant genes responsible for *Arabidopsis thaliana* genes (GA 20-oxidase, *ga1*, *ga1-3*), wheat reduced-height genes (*Rht*) and a rice semidwarf gene (*sd1*) were cloned. These were identified as gibberellin biosynthesis genes or cellular signaling component genes. Stem growth in the mutant background is significantly reduced leading to the dwarf phenotype. Photosynthetic investment in the stem is reduced dramatically as the shorter plants are inherently more stable mechanically. Assimilates become redirected to grain production, amplifying in particular the effect of chemical fertilizers on commercial yield.

HYVs significantly outperform traditional varieties in the presence of adequate irrigation, pesticides, and fertilizers. In the absence of these inputs, traditional varieties may outperform HYVs. Therefore, several authors have challenged the apparent superiority of HYVs not only compared to the traditional varieties alone, but by contrasting the monocultural system associated with HYVs with the polycultural system associated with traditional ones.

Production Increases

Cereal production more than doubled in developing nations between the years 1961–1985. Yields of rice, maize, and wheat increased steadily during that period. The production increases can be attributed roughly equally to irrigation, fertilizer, and seed development, at least in the case of Asian rice.

While agricultural output increased as a result of the Green Revolution, the energy input to produce a crop has increased faster, so that the ratio of crops produced to energy input has decreased over time. Green Revolution techniques also heavily rely on chemical fertilizers, pesticides and herbicides and rely on machines, which as of 2014 rely on or are derived from crude oil, making agriculture increasingly reliant on crude oil extraction. Proponents of the Peak Oil theory fear that a future decline in oil and gas production would lead to a decline in food production or even a Malthusian catastrophe.

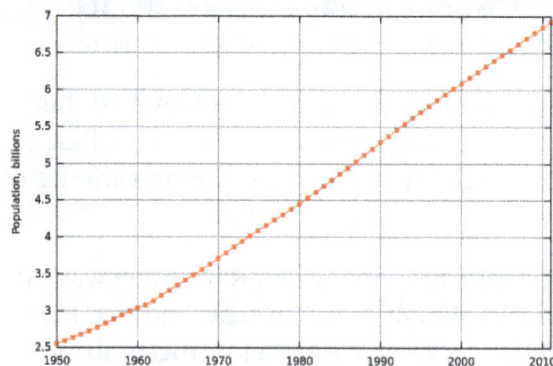

World population 1950–2010

Effects on Food Security

The effects of the Green Revolution on global food security are difficult to assess because of the complexities involved in food systems.

The world population has grown by about four billion since the beginning of the Green Revolution and many believe that, without the Revolution, there would have been greater famine and malnutrition. India saw annual wheat production rise from 10 million tons in the 1960s to 73 million in 2006. The average person in the developing world consumes roughly 25% more calories per day now than before the Green Revolution. Between 1950 and 1984, as the Green Revolution transformed agriculture around the globe, world grain production increased by about 160%.

The production increases fostered by the Green Revolution are often credited with having helped to avoid widespread famine, and for feeding billions of people.

There are also claims that the Green Revolution has decreased food security for a large number of people. One claim involves the shift of subsistence-oriented cropland to cropland oriented towards production of grain for export or animal feed. For example, the Green Revolution replaced much of the land used for pulses that fed Indian peasants for wheat, which did not make up a large portion of the peasant diet.

Criticism

3rd World Economic Sovereignty

A main criticism of the effects of the Green Revolution is the rising costs for many small farmers using HYV seeds, with their associated demands of increased irrigation systems and pesticides. A case study is found in India, where farmers are planting cotton seeds capable of producing Bt toxin. A criticism regarding the Green Revolution are the effects regarding the widespread commercialization and market share of organisations, particularly of the phasing out of seed saving practices in favor of purchasing of seeds, and concerns regarding the financial affordability of the adoption of patented crops amongst farmers, particularly of those in the developing world. This can allow larger farms, even foreign owned farming operations, to buy up local smallhold farms.

Vandana Shiva notes that this is the "second Green Revolution." The first Green Revolution, she notes, was mostly publicly-funded (by the Indian Government). This new Green Revolution, she says, is driven by private [and foreign] interest - notably MNCs like Monsanto. Ultimately, this is leading to foreign ownership over most of India's farmland.

Food Security

Malthusian Criticism

Some criticisms generally involve some variation of the Malthusian principle of population. Such concerns often revolve around the idea that the Green Revolution is unsustainable, and argue that humanity is now in a state of overpopulation or overshoot with regards to the sustainable carrying capacity and ecological demands on the Earth.

Although 36 million people die each year as a direct or indirect result of hunger and poor nutrition, Malthus's more extreme predictions have frequently failed to materialize. In 1798 Thomas Malthus made his prediction of impending famine. The world's population had doubled by 1923 and doubled again by 1973 without fulfilling Malthus's prediction. Malthusian Paul R. Ehrlich, in his 1968 book *The Population Bomb*, said that "India couldn't possibly feed two hundred million more people by 1980" and "Hundreds of millions of people will starve to death in spite of any crash programs." Ehrlich's warnings failed to materialize when India became self-sustaining in cereal production in 1974 (six years later) as a result of the introduction of Norman Borlaug's dwarf wheat varieties.

Since supplies of oil and gas are essential to modern agriculture techniques, a fall in global oil supplies could cause spiking food prices in the coming decades.

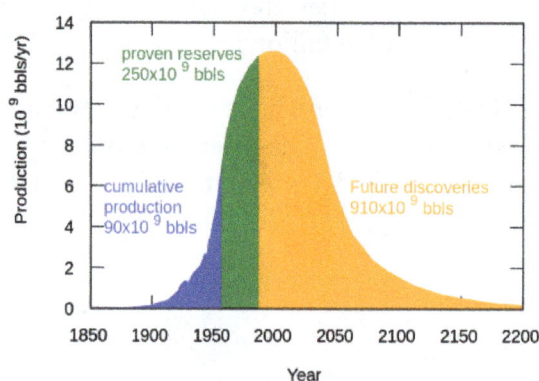

M. King Hubbert's prediction of world petroleum production rates. Modern agriculture is largely reliant on petroleum energy.

Famine

To some modern Western sociologists and writers, increasing food production is not synonymous with increasing food security, and is only part of a larger equation. For example, Harvard professor Amartya Sen claimed large historic famines were not caused by decreases in food supply, but by socioeconomic dynamics and a failure of public action. However, economist Peter Bowbrick disputes Sen's theory, arguing that Sen relies on inconsistent arguments and contradicts available information, including sources that Sen himself cited. Bowbrick further argues that Sen's views coincide with that of the Bengal government at the time of the Bengal famine of 1943, and the policies Sen advocates failed to relieve the famine.

Quality of Diet

Some have challenged the value of the increased food production of Green Revolution agriculture. Miguel A. Altieri, (a pioneer of agroecology and peasant-advocate), writes that the comparison between traditional systems of agriculture and Green Revolution agriculture has been unfair, because Green Revolution agriculture produces monocultures of cereal grains, while traditional agriculture usually incorporates polycultures.

These monoculture crops are often used for export, feed for animals, or conversion into biofuel. According to Emile Frison of Bioversity International, the Green Revolution has also led to a change in dietary habits, as fewer people are affected by hunger and die from starvation, but many are affected by malnutrition such as iron or vitamin-A deficiencies. Frison further asserts that almost 60% of yearly deaths of children under age five in developing countries are related to malnutrition.

High-yield rice (HYR), introduced since 1964 to poverty-ridden Asian countries, such as the Philippines, was found to have inferior flavor and be more glutinous and less savory than their native varieties.This caused its price to be lower than the average market value.

In the Philippines the introduction of heavy pesticides to rice production, in the early part of the Green Revolution, poisoned and killed off fish and weedy green vegetables that traditionally coexisted in rice paddies. These were nutritious food sources for many poor Filipino farmers prior to the introduction of pesticides, further impacting the diets of locals.

Political Impact

A major critic of the Green Revolution, U.S. investigative journalist Mark Dowie, writes:

The primary objective of the program was geopolitical: to provide food for the populace in undeveloped countries and so bring social stability and weaken the fomenting of communist insurgency.

Citing internal Foundation documents, Dowie states that the Ford Foundation had a greater concern than Rockefeller in this area.

There is significant evidence that the Green Revolution weakened socialist movements in many nations. In countries such as India, Mexico, and the Philippines, *technological solutions* were sought as an alternative to expanding *agrarian reform* initiatives, the latter of which were often linked to socialist politics.

Socioeconomic Impacts

The transition from traditional agriculture, in which inputs were generated on-farm, to Green Revolution agriculture, which required the purchase of inputs, led to the widespread establishment of rural credit institutions. Smaller farmers often went into debt, which in many cases results in a loss of their farmland. The increased level of mechanization on larger farms made possible by the Green Revolution removed a large source of employment from the rural economy. Because wealthier farmers had better access to credit and land, the Green Revolution increased class disparities, with the rich–poor gap widening as a result. Because some regions were able to adopt Green Revolution agriculture more readily than others (for political or geographical reasons), interregional economic disparities increased as well. Many small farmers are hurt by the dropping prices resulting from increased production overall. However, large-scale farming companies only account for less than 10% of the total farming capacity. This is a criticism held by many small producers in the food sovereignty movement.

The new economic difficulties of small holder farmers and landless farm workers led to increased rural-urban migration. The increase in food production led to a cheaper food for urban dwellers, and the increase in urban population increased the potential for industrialization.

Globalization

In the most basic sense, the Green Revolution was a product of globalization as evidenced in the creation of international agricultural research centers that shared information, and with transnational funding from groups like the Rockefeller Foundation, Ford Foundation, and United States Agency for International Development (USAID).

Environmental Impact

Increased use of irrigation played a major role in the green revolution.

Biodiversity

The spread of Green Revolution agriculture affected both agricultural biodiversity (or agrodiversity) and wild biodiversity. There is little disagreement that the Green Revolution acted to reduce agricultural biodiversity, as it relied on just a few high-yield varieties of each crop.

This has led to concerns about the susceptibility of a food supply to pathogens that cannot be controlled by agrochemicals, as well as the permanent loss of many valuable genetic traits bred into traditional varieties over thousands of years. To address these concerns, massive seed banks such as Consultative Group on International Agricultural Research's (CGIAR) International Plant Genetic Resources Institute (now Bioversity International) have been established.

There are varying opinions about the effect of the Green Revolution on wild biodiversity. One hypothesis speculates that by increasing production per unit of land area, agriculture will not need to expand into new, uncultivated areas to feed a growing human population. However, land degradation and soil nutrients depletion have forced farmers to clear up formerly forested areas in order to keep up with production. A counter-hypothesis speculates that biodiversity was sacrificed because traditional systems of agriculture that were displaced sometimes incorporated practices to preserve wild biodiversity, and because the Green Revolution expanded agricultural development into new areas where it was once unprofitable or too arid. For example, the development of wheat varieties tolerant to acid soil conditions with high aluminium content, permitted the introduction of agriculture in sensitive Brazilian ecosystems such as Cerrado semi-humid tropical savanna and Amazon rainforest in the geoeconomic macroregions of Centro-Sul and Amazônia. Before the Green Revolution, other Brazilian ecosystems were also significantly damaged by human activity, such as the once 1st or 2nd main contributor to Brazilian megadiversity Atlantic Rainforest (above 85% of deforestation in the 1980s, about 95% after the 2010s) and the important xeric shrublands called Caatinga mainly in Northeastern Brazil (about 40% in the 1980s, about 50% after the 2010s — deforestation of the Caatinga biome is generally associated with greater risks of desertification). This also caused many animal species to suffer due to their damaged habitats.

Nevertheless, the world community has clearly acknowledged the negative aspects of agricultural expansion as the 1992 Rio Treaty, signed by 189 nations, has generated numerous national Biodiversity Action Plans which assign significant biodiversity loss to agriculture's expansion into new domains.

The Green Revolution has been criticized for an agricultural model which relied on a few staple and market profitable crops, and pursing a model which limited the biodiversity of Mexico. One of the critics against these techniques and the Green Revolution as a whole was Carl O. Sauer, a geography professor at the University of California, Berkeley. According to Sauer these techniques of plant breeding would result in negative effects on the country's resources, and the culture:

"A good aggressive bunch of American agronomists and plant breeders could ruin the native resources for good and all by pushing their American commercial stocks... And Mexican agriculture cannot be pointed toward standardization on a few commercial types without upsetting native economy and culture hopelessly... Unless the Americans understand that, they'd better keep out of this country entirely. That must be approached from an appreciation of native economies as being basically sound".

Greenhouse Gas Emissions

According to a study published in 2013 in PNAS, in the absence of the crop germplasm improvement associated with the Green Revolution, greenhouse gas emissions would have been 5.2-7.4 Gt higher than observed in 1965–2004. High yield agriculture has dramatics affects on the amount of carbon cycling in the atmosphere. The way in which farms are grown, in tandem with the seasonal carbon cycling of various crops, could alter the impact carbon in the atmosphere has on global warming. Wheat, rice, and soybean crops, account for a significant amount of the increase n carbon in the atmosphere over the last 50 years.

Dependence on Non-renewable Resources

Most high intensity agricultural production is highly reliant on non-renewable resources. Agricultural machinery and transport, as well as the production of pesticides and nitrates all depend on fossil fuels. Moreover, the essential mineral nutrient phosphorus is often a limiting factor in crop cultivation, while phosphorus mines are rapidly being depleted worldwide. The failure to depart from these non-sustainable agricultural production methods could potentially lead to a large scale collapse of the current system of intensive food production within this century.

Health Impact

The consumption of the pesticides used to kill pests by humans in some cases may be increasing the likelihood of cancer in some of the rural villages using them. Poor farming practices including non-compliance to usage of masks and over-usage of the chemicals compound this situation. In 1989, WHO and UNEP estimated that there were around 1 million human pesticide poisonings annually. Some 20,000 (mostly in developing countries) ended in death, as a result of poor labeling, loose safety standards etc.

Pesticides and Cancer

Long term exposure to pesticides such as organochlorines, creosote, and sulfate have been correlated with higher cancer rates and organochlorines DDT, chlordane, and lindane as tumor promoters in animals. Contradictory epidemiologic studies in humans have linked phenoxy acid herbicides or contaminants in them with soft tissue sarcoma (STS) and malignant lymphoma, organochlorine insecticides with STS, non-Hodgkin's lymphoma (NHL), leukemia, and, less consistently, with cancers of the lung and breast, organophosphorous compounds with NHL and leukemia, and triazine herbicides with ovarian cancer.

Punjab Case

The Indian state of Punjab pioneered green revolution among the other states transforming India into a food-surplus country. The state is witnessing serious consequences of intensive farming using chemicals and pesticide. A comprehensive study conducted by Post Graduate Institute of Medical Education and Research (PGIMER) has underlined the direct relationship between indiscriminate use of these chemicals and increased incidence of cancer in this region. An increase in the number of cancer cases has been reported in several villages including Jhariwala, Koharwala, Puckka, Bhimawali, and Khara.

Environmental activist Vandana Shiva has written extensively about the social, political and economic impacts of the Green Revolution in Punjab. She claims that the Green Revolution's reliance on heavy use of chemical inputs and monocultures has resulted in water scarcity, vulnerability to pests, and incidents of violent conflict and social marginalization.

In 2009, under a Greenpeace Research Laboratories investigation, Dr Reyes Tirado, from the University of Exeter, UK conducted the study in 50 villages in Muktsar, Bathinda and Ludhiana districts revealed chemical, radiation and biological toxicity rampant in Punjab. Twenty percent of the sampled wells showed nitrate levels above the safety limit of 50 mg/l, established by WHO, the study connected it with high use of synthetic nitrogen fertilizers.

Norman Borlaug's Response to Criticism

Borlaug dismissed certain claims of critics, but also cautioned, "There are no miracles in agricultural production. Nor is there such a thing as a miracle variety of wheat, rice, or maize which can serve as an elixir to cure all ills of a stagnant, traditional agriculture."

Of environmental lobbyists, he said:

"some of the environmental lobbyists of the Western nations are the salt of the earth, but many of them are elitists. They've never experienced the physical sensation of hunger. They do their lobbying from comfortable office suites in Washington or Brussels...If they lived just one month amid the misery of the developing world, as I have for fifty years, they'd be crying out for tractors and fertilizer and irrigation canals and be outraged that fashionable elitists back home were trying to deny them these things".

However, the charge of "elitism" could also be leveled against supporters of the Green Revolution. As noted in the documentary Profits from Poison, protective gear for farmers is often designed by people who have never experienced tropical climates. For example, plastic ponchos create a sauna effect if worn in high temperature/humidity, as opposed the experience of wearing them in an airconditioned office.

The New Green Revolution

Although the Green Revolution has been able to improve agricultural output in some regions in the world, there was and is still room for improvement. As a result, many organizations continue to invent new ways to improve the techniques already used in the Green Revolution. Frequently quoted inventions are the System of Rice Intensification, marker-assisted selection, agroecology, and applying existing technologies to agricultural problems of the developing world.

Water Resource Management

Water resource management is the activity of planning, developing, distributing and managing the optimum use of water resources. It is a sub-set of water cycle management. Ideally, water resource management planning has regard to all the competing demands for water and seeks to allocate water on an equitable basis to satisfy all uses and demands. As with other resource management, this is rarely possible in practice.

Overview

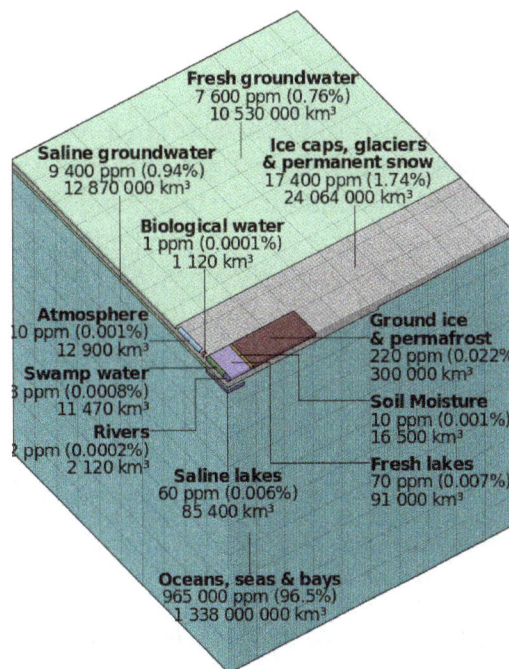

Visualisation of the distribution (by volume) of water on Earth. Each tiny cube (such as the one representing biological water) corresponds to approximately 1000 cubic km of water, with a mass of approximately 1 trillion tonnes (2000 times that of the Great Pyramid of Giza or 5 times that of Lake Kariba, arguably the heaviest man-made object). The entire block comprises 1 million tiny cubes.

Water is an essential resource for all life on the planet. Of the water resources on Earth only three percent of it is fresh and two-thirds of the freshwater is locked up in ice caps and glaciers. Of the remaining one percent, a fifth is in remote, inaccessible areas and much seasonal rainfall in monsoonal deluges and floods cannot easily be used. As time advances, water is becoming scarcer and having access to clean, safe, drinking water is limited among countries. At present only about 0.08 percent of all the world's fresh water is exploited by mankind in ever increasing demand for sanitation, drinking, manufacturing, leisure and agriculture. Due to the small percentage of water remaining, optimizing the fresh water we have left from natural resources has been a continuous difficulty in several locations worldwide.

Much efforts in water resource management is directed at optimizing the use of water and in minimizing the environmental impact of water use on the natural environment. The observation of water as an integral part of the ecosystem is based on integrated water resource management, where the quantity and quality of the ecosystem help to determine the nature of the natural resources.

Successful management of any resources requires accurate knowledge of the resource available, the uses to which it may be put, the competing demands for the resource, measures to and processes to evaluate the significance and worth of competing demands and mechanisms to translate policy decisions into actions on the ground.

For water as a resource this is particularly difficult since sources of water can cross many national boundaries and the uses of water include many that are difficult to assign financial value to and may also be difficult to manage in conventional terms. Examples include rare species or ecosystems or the very long term value of ancient ground water reserves.

Agriculture

Agriculture is the largest user of the world's freshwater resources, consuming 70 percent. As the world population rises it consumes more food (currently exceeding 6%, it is expected to reach 9% by 2050), the industries and urban developments expand, and the emerging biofuel crops trade also demands a share of freshwater resources, water scarcity is becoming an important issue. An assessment of water resource management in agriculture was conducted in 2007 by the International Water Management Institute in Sri Lanka to see if the world had sufficient water to provide food for its growing population or not . It assessed the current availability of water for agriculture on a global scale and mapped out locations suffering from water scarcity. It found that a fifth of the world's people, more than 1.2 billion, live in areas of physical water scarcity, where there is not enough water to meet all their demands. A further 1.6 billion people live in areas experiencing economic water scarcity, where the lack of investment in water or insufficient human capacity make it impossible for authorities to satisfy the demand for water.

The report found that it would be possible to produce the food required in future, but that continuation of today's food production and environmental trends would lead to crises in many parts of the world. Regarding food production, the World Bank targets agricultural food production and water resource management as an increasingly global issue that is fostering an important and growing debate. The authors of the book *Out of Water: From abundance to Scarcity and How to Solve the World's Water Problems*, which laid down a six-point plan for solving the world's water problems. These are: 1) Improve data related to water; 2) Treasure the environment; 3) Reform water governance; 4) Revitalize agricultural water use; 5) Manage urban and industrial demand; and 6) Empower the poor and women in water resource management. To avoid a global water crisis, farmers will have to strive to increase productivity to meet growing demands for food, while industry and cities find ways to use water more efficiently.

Managing Water in Urban Settings

As the carrying capacity of the Earth increases greatly due to technological advances, urbanization in modern times occurs because of economic opportunity. This rapid urbanization happens worldwide but mostly in new rising economies and developing countries. Cities in Africa and Asia are growing fastest with 28 out of 39 megacities (a city or urban area with more than 10 million inhabitants) worldwide in these developing nations. The number of megacities will continue to rise reaching approximately 50 in 2025. With developing economies water scarcity is a very common and very prevalent issue. Global freshwater resources dwindle in the eastern hemisphere either than at the poles, and with the majority of urban development millions live with insufficient fresh water. This is caused by polluted freshwater resources, overexploited groundwater resources, insufficient harvesting capacities in the surrounding rural areas, poorly constructed and maintained water supply systems, high amount of informal water use and insufficient technical and water management capacities.

In the areas surrounding urban centres, agriculture must compete with industry and municipal users for safe water supplies, while traditional water sources are becoming polluted with urban wastewater. As cities offer the best opportunities for selling produce, farmers often have no alternative to using polluted water to irrigate their crops. Depending on how developed a city's wastewater treatment is, there can be significant health hazards related to the use of this water. Wastewater from cities can contain a mixture of pollutants. There is usually wastewater from kitchens and toilets along with rainwater runoff. This means that the water usually contains excessive levels of nutrients and salts, as well as a wide range of pathogens. Heavy metals may also be present, along with traces of antibiotics and endocrine disruptors, such as oestrogens.

Developing world countries tend to have the lowest levels of wastewater treatment. Often, the water that farmers use for irrigating crops is contaminated with pathogens from sewage. The pathogens of most concern are bacteria, viruses and parasitic worms, which directly affect farmers' health and indirectly affect consumers if they eat the contaminated crops. Common illnesses include diarrhoea, which kills 1.1 million people annually and is the second most common cause of infant deaths. Many cholera outbreaks are also related to the reuse of poorly treated wastewater. Actions that reduce or remove contamination, therefore, have the potential to save a large number of lives and improve livelihoods. Scientists have been working to find ways to reduce contamination of food using a method called the 'multiple-barrier approach'.

This involves analysing the food production process from growing crops to selling them in markets and eating them, then considering where it might be possible to create a barrier against contamination. Barriers include: introducing safer irrigation practices; promoting on-farm wastewater treatment; taking actions that cause pathogens to die off; and effectively washing crops after harvest in markets and restaurants.

Future of Water Resources

One of the biggest concerns for our water-based resources in the future is the sustainability of the current and even future water resource allocation. As water becomes more scarce, the importance of how it is managed grows vastly. Finding a balance between what is needed by humans and what is needed in the environment is an important step in the sustainability of water resources. Attempts to create sustainable freshwater systems have been seen on a national level in countries such as Australia, and such commitment to the environment could set a model for the rest of the world.

The field of water resources management will have to continue to adapt to the current and future issues facing the allocation of water. With the growing uncertainties of global climate change and the long term impacts of management actions,the decision-making will be even more difficult. It is likely that ongoing climate change will lead to situations that have not been encountered. As a result, new management strategies will have to be implemented in order to avoid setbacks in the allocation of water resources.

Nutrient Management

Nitrogen fertilizer being applied to growing corn (maize) in a contoured, no-tilled field in Iowa.

Nutrient management is the science and art directed to link soil, crop, weather, and hydrologic factors with cultural, irrigation, and soil and water conservation practices to achieve the goals of optimizing nutrient use efficiency, yields, crop quality, and economic returns, while reducing off-site transport of nutrients that may impact the environment. Nutrient management is the skillful task of matching a specific field soil, climate, and crop management conditions to rate, source, timing, and place (commonly known as the 4R nutrient stewardship) of nutrient application.

Some important factors that need to be considered when managing nutrients include (a) the application of nutrients considering the achievable optimum yields and, in some cases, crop quality; (b) the management, application, and timing of nutrients using a budget based on all sources and sinks active at the site; and (c) the management of soil, water, and crop to minimize the off-site transport of nutrients from nutrient leaching out of the root zone, surface runoff, and volatilization (or other gas exchanges).

There can be potential interactions because of differences in nutrient pathways and dynamics. For instance, practices that reduce the off-site surface transport of a given nutrient may increase the leaching losses of other nutrients. These complex dynamics present nutrient managers the difficult task of integrating soil, crop, weather, hydrology, and management practices to achieve the best balance for maximizing profit while contributing to the conservation of our biosphere.

Nutrient Management Plan

Manure spreader

A crop nutrient management plan is a tool that farmers can use to increase the efficiency of all the nutrient sources a crop uses while reducing production and environmental risk, ultimately increasing profit. It is generally agreed that there are ten fundamental components of a Crop Nutrient Management Plan. Each component is critical to helping analyze each field and improve nutrient efficiency for the crops grown. These components include:

- *Field Map:* The map, including general reference points (such as streams, residences, wellheads etc.), number of acres, and soil types is the base for the rest of the plan.

- *Soil Test:* How much of each nutrient (N-P-K and other critical elements such as pH and organic matter) is in the soil profile? The soil test is a key component needed for developing the nutrient rate recommendation.

- *Crop Sequence:* Did the crop that grew in the field last year (and in many cases two or more years ago) fix nitrogen for use in the following years? Has long-term no-till increased organic matter? Did the end-of-season stalk test show a nutrient deficiency? These factors also need to be factored into your plan.

- *Estimated Yield:* Factors that affect yield are numerous and complex. A field's soils, drainage, insect, weed and disease pressure, rotation and many other factors differentiate one field from another. This is why using historic yields is important in developing yield estimates for next year. Accurate yield estimates can dramatically improve nutrient use efficiency.

- *Sources and Forms:* The sources and forms of available nutrients can vary from farm-to-farm and even field-to-field. For instance, manure fertility analysis, storage practices and other factors will need to be included in a nutrient management plan. Manure nutrient tests/analysis are one way to determine the fertility of it. Nitrogen fixed from a previous year's legume crop and residual affects of manure also effects rate recommendations. Many other nutrient sources should also be factored into this plan.

- *Sensitive Areas:* What's out of the ordinary about a field's plan? Is it irrigated? Next to a stream or lake? Especially sandy in one area? Steep slope or low area? Manure applied in one area for generations due to proximity of dairy barn? Extremely productive—or unproductive—in a portion of the field? Are there buffers that protect streams, drainage ditches, wellheads, and other water collection points? How far away are the neighbors? What's the general wind direction? This is the place to note these and other special conditions that need to be considered.

- *Recommended rates:* Here's the place where science, technology, and art meet. Given everything you've noted, what is the optimum rate of N, P, K, lime and any other nutrients? While science tells us that a crop has changing nutrient requirements during the growing season, a combination of technology and farmer's management skills assure optimum nutrient availability at all stages of growth. No-till corn generally requires starter fertilizer to give the seedling a healthy start.

- *Recommended timing:* When does the soil temperature drop below 50 degrees? Will a N stabilizer be used? What's the tillage practice? Strip-till corn and no-till often require different timing approaches than seed planted into a field that's been tilled once with a field cultivator. Will a starter fertilizer be used to give the seedling a healthy start? How many acres can be covered with available labor (custom or hired) and equipment? Does manure application in a farm depend on a custom applicator's schedule? What agreements have been worked out with neighbors for manure use on their fields? Is a neighbor hosting a special event? All these factors and more will likely figure into the recommended timing.

- *Recommended methods:* Surface or injected? While injection is clearly preferred, there may be situations where injection is not feasible (i.e. pasture, grassland). Slope, rainfall patterns, soil type, crop rotation and many other factors determine which method is best for optimizing nutrient efficiency (availability and loss) in farms. The combination that's right in one field may differ in another field even with the same crop.

- *Annual review and update:* Even the best managers are forced to deviate from their plans. What rate was actually applied? Where? Using which method? Did an unusually mild winter or wet spring reduce soil nitrate? Did a dry summer, disease, or some other unusual factor increase nutrient carryover? These and other factors should be noted. It's easier to make notes throughout the year than to remember back six to 10 months.

When such a plan is designed for animal feeding operations (AFO), it may be termed a "manure management plan." In the United States, some regulatory agencies recommend or require that farms implement these plans in order to prevent water pollution. The U.S. Natural Resources Conservation Service (NRCS) has published guidance documents on preparing a comprehensive nutrient management plan (CNMP) for AFOs.

The International Plant Nutrition Institute has published a 4R plant nutrition manual for improving the management of plant nutrition. The manual outlines the scientific principles behind each of the four R's or "rights" and discusses the adoption of 4R practices on the farm, approaches to nutrient management planning, and measurement of sustainability performance.

Agricultural Wastewater Treatment

Agricultural wastewater treatment is a farm management agenda for controlling pollution from surface runoff that may be contaminated by chemicals in fertiliser, pesticides, animal slurry, crop residues or irrigation water.

Nonpoint Source Pollution

Nonpoint source pollution from farms is caused by surface runoff from fields during rain storms. Agricultural runoff is a major source of pollution, in some cases the only source, in many watersheds.

Sediment Runoff

Highly erodible soils on a farm in Iowa

Soil washed off fields is the largest source of agricultural pollution in the United States. Excess sediment causes high levels of turbidity in water bodies, which can inhibit growth of aquatic plants, clog fish gills and smother animal larvae.

Farmers may utilize erosion controls to reduce runoff flows and retain soil on their fields. Common techniques include:

- contour ploughing
- crop mulching
- crop rotation
- planting perennial crops
- installing riparian buffers.

Nutrient Runoff

Nitrogen and phosphorus are key pollutants found in runoff, and they are applied to farmland in several ways, such as in the form of commercial fertilizer, animal manure, or municipal or industrial wastewater (effluent) or sludge. These chemicals may also enter runoff from crop residues, irrigation water, wildlife, and atmospheric deposition.

Manure spreader

Farmers can develop and implement nutrient management plans to mitigate impacts on water quality by:

- mapping and documenting fields, crop types, soil types, water bodies

- developing realistic crop yield projections

- conducting soil tests and nutrient analyses of manures and/or sludges applied

- identifying other significant nutrient sources (e.g., irrigation water)

- evaluating significant field features such as highly erodible soils, subsurface drains, and shallow aquifers

- applying fertilizers, manures, and/or sludges based on realistic yield goals and using precision agriculture techniques.

Pesticides

Aerial application (crop dusting) of pesticides over a soybean field in the U.S.

Pesticides are widely used by farmers to control plant pests and enhance production, but chemical pesticides can also cause water quality problems. Pesticides may appear in surface water due to:

- direct application (e.g. aerial spraying or broadcasting over water bodies)

- runoff during rain storms

- aerial drift (from adjacent fields).

Some pesticides have also been detected in groundwater.

Farmers may use Integrated Pest Management (IPM) techniques (which can include biological pest control) to maintain control over pests, reduce reliance on chemical pesticides, and protect water quality.

There are few safe ways of disposing of pesticide surpluses other than through containment in well managed landfills or by incineration. In some parts of the world, spraying on land is a permitted method of disposal.

Point Source Pollution

Farms with large livestock and poultry operations, such as factory farms, can be a major source of point source wastewater. In the United States, these facilities are called *concentrated animal feeding operations* or *confined animal feeding operations* and are being subject to increasing government regulation.

Animal Wastes

Confined Animal Feeding Operation in the United States

The constituents of animal wastewater typically contain

- Strong organic content — much stronger than human sewage

- High solids concentration

- High nitrate and phosphorus content

- Antibiotics

- Synthetic hormones

- Often high concentrations of parasites and their eggs

- Spores of *Cryptosporidium* (a protozoan) resistant to drinking water treatment processes

- Spores of *Giardia*

- Human pathogenic bacteria such as *Brucella* and *Salmonella*

Animal wastes from cattle can be produced as solid or semisolid manure or as a liquid slurry. The production of slurry is especially common in housed dairy cattle.

Treatment

Whilst solid manure heaps outdoors can give rise to polluting wastewaters from runoff, this type of waste is usually relatively easy to treat by containment and/or covering of the heap.

Animal slurries require special handling and are usually treated by containment in lagoons before disposal by spray or trickle application to grassland. Constructed wetlands are sometimes used to facilitate treatment of animal wastes, as are anaerobic lagoons. Excessive application or application to sodden land or insufficient land area can result in direct runoff to watercourses, with the potential for causing severe pollution. Application of slurries to land overlying aquifers can result in direct contamination or, more commonly, elevation of nitrogen levels as nitrite or nitrate.

The disposal of any wastewater containing animal waste upstream of a drinking water intake can pose serious health problems to those drinking the water because of the highly resistant spores present in many animals that are capable of causing disabling disease in humans. This risk exists even for very low-level seepage via shallow surface drains or from rainfall run-off.

Some animal slurries are treated by mixing with straws and composted at high temperature to produce a bacteriologically sterile and friable manure for soil improvement.

Piggery Waste

Hog confinement barn or piggery

Piggery waste is comparable to other animal wastes and is processed as for general animal waste, except that many piggery wastes contain elevated levels of copper that can be toxic in the natural environment. The liquid fraction of the waste is frequently separated off and re-used in the piggery to avoid the prohibitively expensive costs of disposing of copper-rich liquid. Ascarid worms and their eggs are also common in piggery waste and can infect humans if wastewater treatment is ineffective.

Silage Liquor

Fresh or wilted grass or other green crops can be made into a semi-fermented product called silage which can be stored and used as winter forage for cattle and sheep. The production of silage often involves the use of an acid conditioner such as sulfuric acid or formic acid. The process of silage making frequently produces a yellow-brown strongly smelling liquid which is very rich in simple sugars, alcohol, short-chain organic acids and silage conditioner. This liquor is one of the most polluting organic substances known. The volume of silage liquor produced is generally in proportion to the moisture content of the ensiled material.

Treatment Silage liquor is best treated through prevention by wilting crops well before silage making. Any silage liquor that is produced can be used as part of the food for pigs. The most effective treatment is by containment in a slurry lagoon and by subsequent spreading on land following substantial dilution with slurry. Containment of silage liquor on its own can cause structural problems in concrete pits because of the acidic nature of silage liquor.

Milking parlour (Dairy Farming) Wastes

Although milk has a deserved reputation as an important and valuable food product, its presence in wastewaters is highly polluting because of its organic strength, which can lead to very rapid de-oxygenation of receiving waters. Milking parlour wastes also contain large volumes of wash-down water, some animal waste together with cleaning and disinfection chemicals.

Treatment Milking parlour wastes are often treated in admixture with human sewage in a local sewage treatment plant. This ensures that disinfectants and cleaning agents are sufficiently diluted and amenable to treatment. Running milking wastewaters into a farm slurry lagoon is a possible option although this tends to consume lagoon capacity very quickly. Land spreading is also a treatment option.

Slaughtering Waste

Wastewater from slaughtering activities is similar to milking parlour waste although considerably stronger in its organic composition and therefore potentially much more polluting.

Treatment As for milking parlour waste.

Vegetable Washing Water

Washing of vegetables produces large volumes of water contaminated by soil and vegetable pieces. Low levels of pesticides used to treat the vegetables may also be present together with moderate levels of disinfectants such as chlorine.

Treatment Most vegetable washing waters are extensively recycled with the solids removed by settlement and filtration. The recovered soil can be returned to the land.

Firewater

Although few farms plan for fires, fires are nevertheless more common on farms than on many other industrial premises. Stores of pesticides, herbicides, fuel oil for farm machinery and fertilizers can all help promote fire and can all be present in environmentally lethal quantities in firewater from fire fighting at farms.

Treatment All farm environmental management plans should allow for containment of substantial quantities of firewater and for its subsequent recovery and disposal by specialist disposal companies. The concentration and mixture of contaminants in firewater make them unsuited to any treatment method available on the farm. Even land spreading has produced severe taste and odour problems for downstream water supply companies in the past.

Ecological Sanitation

Ecosan concept showing a separation of flow streams, treatment and reuse

Ecological sanitation, commonly abbreviated to ecosan (also spelled eco-san or EcoSan), is an approach which is characterized by a desire to safely "close the loop" (mainly for the nutrients and organic matter) between sanitation and agriculture. Ecosan systems safely recycle excreta resources (plant nutrients and organic matter) to crop production in such a way that the use of non-renewable resources is minimised. When properly designed and operated, ecosan systems can strive to provide a hygienically safe, economical, and closed-loop system to convert human excreta into nutrients to be returned to the soil, and water to be returned to the land.

Definition

The definitions of ecosan have varied in the past. Nowadays, the most widely accepted definition of ecosan, formulated by experts in Sweden, is: "Ecological sanitation systems are systems which allow for the safe recycling of nutrients to crop production in such a way that the use of non-renewable resources is minimized. These systems have a strong potential to be sustainable sanitation systems if technical, institutional, social and economic aspects are managed appropriately."

Traditionally, ecosan has often been associated with urine diversion and in particular with urine-diverting dry toilets (UDDTs), a type of dry toilet. For this reason, the term "ecosan toilet" is widely used when people mean a UDDT. However, the ecosan concept should not be limited to one particular type of toilet. Also, UDDTs can be used without having any reuse activities in which case they are not in line with the ecosan concept (an example being the 80,000 UDDTs implemented by eThekwini Municipality near Durban, South Africa).

Overview

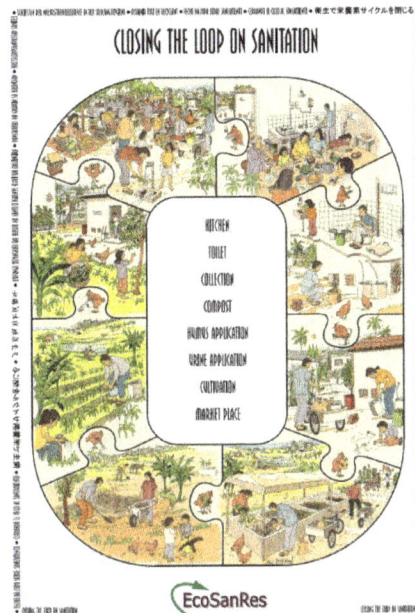

Poster by EcoSanRes program: Closing the loop on Sanitation (2005)

Ecosan closing the loop poster (in French), by the NGO CREPA in 2005, UDDTs are used in this example

Objectives

The main objectives of ecological sanitation are to reduce the health risks related to sanitation, contaminated water and waste; to prevent groundwater pollution and surface water pollution; and to reuse nutrients or energy contained within wastes.

Resource Recovery

The statement in the definition of ecosan to "safely recycle" includes hygienic, microbial and chemical aspects. Thus, the recycled human excreta product, in solid or liquid form, shall be of high quality both concerning pathogens and all kind of hazardous chemical components. The statement "use of non-renewable resources is minimised" means that the gain in resources by recycling shall be larger than the cost of resources by recycling.

Ecosan is based on an overall concept of material flows as part of an ecologically and economically sustainable wastewater management system tailored to the needs of the users and to the respective local conditions. It does not favour a specific sanitation technology, but is rather a certain philosophy in handling substances that have so far been seen simply as wastewater and water-carried waste for disposal.

Reuse as Fertilizer

Right from the start, the first proponents of ecosan systems had a strong focus on increasing agricultural productivity (via the reuse of excreta as fertilizers) and thus improving the nutritional status of the people at the same time as providing them with safe sanitation. Disease reduction was meant to be achieved not only by reducing infections transmitted via the fecal-oral route but also by reducing malnutrition in children. This link between WASH, nutrition, a disease called environmental enteropathy (or tropical enteropathy) as well as stunted growth of children has risen to the top of the agenda of the WASH sector since about 2013.

Agricultural trials around the world have shown measurable benefits of using treated excreta in agriculture as a fertilizer and soil conditioner. This applies in particular to the use of urine. Reuse trials in Zimbabwe showed positive results for using urine on green, leafy plants such as spinach or maize as well as fruit trees. Another study in Finland indicated that the use of urine and the use of urine and wood ash "could produce 27% and 10% more red beet root biomass." Urine has been proven in many studies to be a valuable, relatively easy to handle fertilizer, containing nitrogen, phosphorus, potassium and important micro-nutrients.

Phosphorus Recovery

Another aspect that ecosan systems are trying to address is the possible upcoming shortage of phosphorus. Phosphorus has an important role for plant growth, and therefore in fertilizer production, but is a limited mineral resource. The situation is similar for potassium. Known mineral phosphate rock reserves are becoming scarce and increasingly costly to extract - this is also called the "peak phosphorus" crisis. One review of global phosphate supply suggested that if collected, phosphate in urine could supply 22% of the total demand.

Comparison with Sustainable Sanitation

The definition of ecosan is focusing on the health, environment and resource aspect of sustainable sanitation. Thus ecosan is not, per se, sustainable sanitation, but ecosan systems can be implemented in a sustainable way and have a strong potential for sustainable sanitation, if technical, institutional, social and economical aspects are cared for appropriately. Ecosan systems can be "unsustainable" for example if there is too little user acceptance or if the costs of the system are too high for a given target group of users, making the system financially unsustainable in the longer term.

Possible Benefits

Benefits of ecosan systems may include:

- Minimizing the introduction of pathogens from human excreta into the water cycle (groundwater and surface water) - a major consideration in low-lying geographies is groundwater pollution by pit latrines. In many areas where the water table is high, pit latrines directly pollute the water table, potentially affecting the large numbers of people.

- Promotion of safe, hygienic recovery and use of nutrients (nitrogen and phosphorus), organics, trace elements, water and energy

- Promotion of soil fertility, improvement of agricultural productivity and food security

- Contribution to the conservation of resources through lower water consumption, substitution of mineral fertilizer and minimization of water pollution

- Less reliance on mined phosphorus for fertilizer production

- Energy reduction in fertilizer production: Urea is the major component of urine, yet we produce vast quantities of urea by using fossil fuels. By properly managing urine, treatment costs as well as fertilizer costs can be reduced.

Challenges

The ecosan approach has been criticised for being overly focused on reuse in agriculture, whilst neglecting some of the other criteria for sustainable sanitation. In fact, ecosan systems can be "unsustainable" for example if there is too little user acceptance or if the costs of the system are too high for a given target group of users, making the system financially unsustainable in the longer term.

Some proponents of ecosan have been criticised as being too dogmatic, with an over-emphasis on environmental resource protection rather than a focus on public health protection and provision of sanitation at a very low cost (for example UDDTs, which some people call "ecosan toilets" may be more expensive to build than pit latrines, even if in the longer term they are cheaper to maintain).

The safety of ecosan systems in terms of pathogen destruction during the various treatment processes is a continuous topic of debate between proponents and opponents of ecosan systems. However, the publication of the WHO Guidelines on Reuse, with its multiple barrier concept, has gone a long way in establishing a common framework for safe reuse. Nevertheless, the question remains whether ecosan systems can ever be scaled up to reach millions of people and how they can be made sufficiently safe to operate. The initial excitement in the early 2000s by the ecosan pioneers has changed into a realization that changing attitudes and behaviours in sanitation takes a lot of patience.

Acknowledgement for ecosan came with the awarding of the Stockholm Water Prize in 2013 to Peter Morgan, a pioneer of handpumps and ventilated pit latrines (VIPs) in addition to ecosan-type toilets (the Arborloo, the Skyloo and the Fossa alterna). Peter Morgan is renowned as one of the leading creators and proponents of ecological sanitation solutions, which enable the safe reuse of human excreta to enhance soil quality and crop production. His ecosan-type toilets are now in use in countries across the globe, centred on converting a sanitary problem into a productive resource.

Also many of the research projects that the Bill and Melinda Gates Foundation have been funding since about 2011 in sanitation are dealing with resource recovery - this might well be a legacy of the ecosan concept, even if the term "ecosan" is not used by these researchers.

Technologies Used in Ecosan Systems

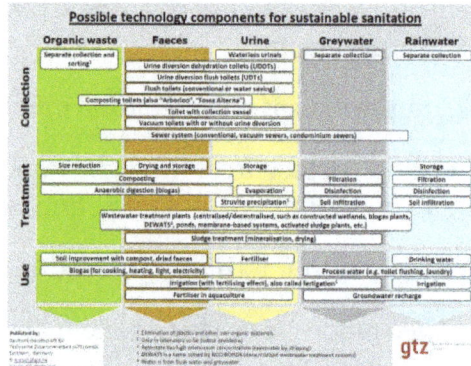

Possible technology components for sustainable sanitation, of which ecosan is a sub-set focussing on the reuse possibilities

Ecosan offers a flexible framework, where centralised elements can be combined with decentralised ones, waterborne with dry sanitation, high-tech with low-tech, etc. By considering a much larger range of options, optimal and economic solutions can be developed for each particular situation. Technologies used in ecosan systems often - but not always - include elements of source separation, i.e. keeping different waste streams separate, as this can make treatment and safe reuse easier.

The most common technology used in ecosan systems is the urine-diverting dry toilet, but ecosan systems can also use other technologies, such as vacuum toilets coupled with biogas plants, constructed wetlands, composting toilets and so forth.

Examples of ecosan projects worldwide can be found in a list published by GIZ in 2012, as well as in those case studies published by the Sustainable Sanitation Alliance that are focused on reuse activities.

History

Excreta Reuse in Dry Sanitation Systems

The recovery and use of urine and feces in "dry sanitation systems", i.e. without sewers or without mixing substantial amounts of water with the excreta, has been practiced by almost all cultures. The reuse was not limited to agricultural production. The Romans, for example, were aware of the bleaching attribute of the ammonia within urine and used it to whiten clothing.

Many traditional agricultural societies recognized the value of human waste for soil fertility and practised the "dry" collection and reuse of excreta. This enabled them to live in communities in which nutrients and organic matter contained in excreta were returned to the soil. Historical descriptions about these practices are sparse, but it is known that excreta reuse was practiced widely in Asia (for example in China, Japan, Vietnam, Cambodia, Korea) but also in Central and South America. However, the most renowned example of the organised collection and use of human excreta to support food production is that of China. The value of "night soil" as a fertilizer was recognized with well-developed systems in place to enable the collection of excreta from cities and its transportation to fields. The Chinese were aware of the benefits of using excreta in crop production more than 2500 years ago, enabling them to sustain more people at a higher density than any other system of agriculture.

In Mexico the Aztec culture collected human excreta for agricultural use. One example of this practice has been documented for the Aztec city of Tenochtitlan which was founded in 1325 and was one of the last cities of pre-Hispanic Mexico (conquered in 1521 by the Spanish): The population placed the sweepings in special boats moored at docks around the city. Mixtures of sweepings and excreta were used to fertilize the chinampas (agricultural fields) or to bolster the banks bordering the lake. Urine was collected in containers in all houses, then mixed with mud and used as a fabric dye. The Aztecs recognized the importance of recycling nutrients and compounds contained in wastewater.

In Peru, the Incas had a high regard for excreta as a fertilizer, which was stored, dried and pulverized to be utilized when planting maize.

In the Middle Ages, the use of excreta and greywater in agricultural production was the norm. European cities were rapidly urbanizing and sanitation was becoming an increasingly serious problem, whilst at the same time the cities themselves were becoming an increasingly important source of agricultural nutrients. The practice of directly using the nutrients in excreta and wastewater for agriculture therefore continued in Europe into the middle of the 19th Century. Farmers, recognizing the value of excreta, were eager to get these fertilizers to increase production and urban sanitation benefited. This practice was also called gong farmer in England but carried many health risks for those involved with transporting the excreta and fecal sludge.

Traditional forms of sanitation and excreta reuse have continued in various parts of the world for centuries and were still common practice at the advent of the Industrial Revolution. Even as the world became increasingly more urbanised, the nutrients in excreta collected from urban sanitation systems without mixing with water were still used in many societies as a resource to maintain soil fertility, despite rising population densities.

Decline in Recovery of Nutrients from Human Excreta in Dry Systems

Recovery of nutrients and organic matter from excreta and greywater in non-sewered sanitation systems was addressing the sanitation problems in settlements in Europe and elsewhere and was contributing to securing agricultural productivity. However, the practice did not become the dominant approach to urban sanitation in the 20th century and was gradually replaced with sewer-based sanitation systems without nutrient recovery (apart from agricultural reuse of sewage sludge in some cases) - at least for cities that can afford it. There were four main driving factors that led to the demise in the recovery and use of excreta and greywater from European cities in the 19th century:

- Growth of urban settlements and increasing distance from agricultural fields: urban settlements had grown dramatically over the centuries. The logistical challenge of removing the feces of a booming population from densely packed city centres to bring to agricultural areas many miles away proved too great.

- Increasing water consumption and use of flush toilet: Water flushing greatly increased the volume of sewage, at the same time diluting the nutrients, making it virtually impossible for them to be recovered and reused as they previously were.

- Production of cheap synthetic fertilizers: The nutrient demand of farmers was eventually met by cheap chemical fertilizers, making any efforts to recover and reuse the nutrients and organic material from the large volumes of sewage completely obsolete.

- Political intervention as a consequence of the perceived need for a change with regards to how to deal with odorous substances: Up until the end of the nineteenth century the dominant theory on the spread of illness was the miasma theory. This theory stated that everything that smelled had to be gotten rid off because inhaling bad smells was thought to lead to illness. The miasma theory did contribute to containing disease in urban settlements, but did not allow for a suitable approach to safe excreta reuse to be adopted.

Interestingly, the use of (odorous) animal manure in agriculture continued through to this day, probably because the odour of manure was not thought to contribute to human illnesses and because the vast amount of manure could not be "flushed away" like human excreta can.

Even in situations where dry sanitation systems have been replaced by sewer-based sanitation systems, the recovery of nutrients from wastewater may continue in two forms:

- Wastewater reuse or resource recovery: Use of raw, treated or partially treated wastewater for irrigation in agriculture (with the associated health risks if it is done in an improper way which is often the case in developing countries); and

- Application of sewage sludge to agricultural lands which is not without controversy in many industrialised countries due to the risks of polluting soils with heavy metals and micropollutants if not managed properly.

Research from 1990s onwards

The Swedish International Development Cooperation Agency (Sida) funded the "SanRes R&D programme" during 1993 to 2001 which lay the foundation for the subsequent "EcoSanRes programme" carried out by Stockholm Environment Institute (2002–2011). A publication by Sida called "Ecological sanitation" in 1998 compiled the knowledge generated to date about ecosan in a popular book which was published as a second edition in 2004. The book has also been translated into Chinese, French and Spanish.

The German government enterprise GIZ also had a large "ecosan program" from 2001 to 2012. Whilst the term "ecosan" was preferred in the initial stages of this program, it was from 2007 onwards more and more replaced by the broader term "sustainable sanitation". In fact, the Sustainable Sanitation Alliance was founded in 2007 in an attempt to broaden the ecosan concept and to bring together various actors under one umbrella.

Research into how to make reuse of urine and feces safe in agriculture was carried out by Swedish researchers, for example Hakan Jönsson and his team, whose publication on "Guidelines on the Use of Urine and feces in Crop Production" was a milestone which was later incorporated into the WHO "Guidelines on Safe Reuse of Wastewater, Excreta and Greywater" from 2006. The multiple barrier concept to reuse, which is the key cornerstone of this publication, has led to a clear understanding on how excreta reuse can be done safely.

Workshops and Conferences

Initially, there were dedicated "ecosan conferences" to present and discuss research on ecosan projects:

- A first workshop on ecological sanitation was held in Balingsholm, Sweden in 1997, where all the then established ecosan experts, such as Håkan Jönsson, Peter Morgan (winner of the 2013 Stockholm Water Prize), Ron Sawyer, George Anna Clark and Gunder Edström participated.

- Workshop in Mexico in 1999 with the title "Closing the Loop - Ecological sanitation for food security"

- Ecosan conference in Bonn, Germany in 2000

- First international ecosan conference in Nanning, China in 2001

- Second ecosan conference in Lübeck Germany in 2003

- Third ecosan conference in Durban, South Africa in 2005

- Ecosan conference in Fortaleza, Brazil called "International Conference on Sustainable Sanitation - Water and Food Security for Latin America" in 2007

Since then the ecosan theme has been integrated into other WASH conferences, and separate large ecosan conferences have no longer been organised.

Use of the Term "Ecosan"

The term "ecosan" was first used in about the 1990s (or perhaps even late 1980s) by an NGO in Ethiopia called Sudea. They used it for urine-diverting dry toilets coupled with reuse activities.

In the ecosan concept, human excreta and wastewater is regarded as a potential resource - which is why it has also been called "resource oriented sanitation". The term "productive sanitation" has also been in use since about 2006.

Disputes Amongst Experts

In the early days during the 1990s when the term ecosan was something new, discussions were heated and confrontational. Supporters of ecosan claimed the corner on containment, treatment and reuse. The "establishment" defended deep pit latrines, and waterborne systems and they felt offended. Ecosan supporters criticised conventional sanitation for killing children, contaminating waterways with nutrients and pathogens and making helminth worms a global pandemic with upwards of 2 billion cases. Since then, the two opposing sides have slowly found ways of dealing with each other, and the formation of the Sustainable Sanitation Alliance has further helped to provide a space for all sanitation actors to meet and push into the same direction of sustainable sanitation.

Examples

- Sweden is the leader in Europe to put ecosan into practice at a larger scale. For example, Tanum Municipality in Sweden has introduced urine separation toilets due to their very rocky and challenging terrain initially, and later also to recover phosphorus.

- Sweden has also made it possible in 2013 to certify safe and sanitized blackwater (urine and human excreta) from blackwater systems and for further use as a recognized fertilizer. Such blackwater systems could be vacuum toilets or septic tanks. The criteria for the certification have been developed by the Swedish Institute for Agricultural and Environmental Engineering and may pave the way for farmers to use human waste for agricultural production. The Federation of Swedish Farmers have been active in this development. Furthermore, the Swedish EPA in their last proposal in 2014 has downgraded the hygiene risk associated with urine. Previously the normal storage requirement for hygienic quality for large scale use of urine was 6 months. Now they propose decreasing this to one month.

- Stockholm Environment Institute (SEI) ran a large worldwide ecosan research programme called "Ecosanres" from 2001 to 2011. One of the "dry ecosan" pilot projects (i.e. with using dry toilets) of this programme was a large scale implementation of urine-diverting dry toilets (UDDTs) in multi-story buildings together with other technologies to allow resource recovery from excreta. This project was called the Erdos Eco-Town Project in a town called Erdos in the Inner Mongolia Autonomous Region of China. It was a collaboration between the Dongsheng District government in Erdos and the Stockholm Environment Institute and aimed to save water and provide sanitation services in this drought-stricken and rapidly urbanizing area of northern China. For a variety of technical, social and institutional reasons, the UDDTs were removed after only a few years and the project failed to deliver in the area of nutrient recovery. This project is now well documented and has raised more awareness of the challenges and disadvantages of "urban ecosan".

- SOIL in Haiti built what they call "ecosan toilets" (UDDTs) as part of the emergency relief effort following the 2010 Haiti earthquake. More than 20,000 Haitians are currently using SOIL ecological sanitation toilets and SOIL has produced over 400,000 liters of compost as a result. The compost is used for agricultural and reforestation projects. SOIL's composting process is effective in inactivating *Ascaris* eggs - an indicator for helminth eggs in general - in the excreta collected from the dry toilets within 16 weeks. The composting and monitoring methos used by SOIL in Haiti may serve as an example for other international settings.

- Wherever the Need, an NGO in the UK build ecosan facilities (UDDTs) in various parts of the developing world. They predominantly work in Tamil Nadu (India), where the Tamil Nadu

State Government provides subsidies for their work. Wherever the Need have also constructed ecosan in other parts of rural India, Kenya and Sierra Leone. According to their website, heir ecosan projects have positively affected 50,000 people in the developing world.

- The NGO CREPA which was operating in the French-speaking West Africa region (now bi-lingual and called WSA - Water and Sanitation in Africa) was very active in ecosan promotion during the years of 2002 - 2010 with a strong focus on UDDTs coupled with reuse in agriculture, especially in Burkina Faso.

References

- Kunstler, James Howard (2012). Too Much Magic; Wishful Thinking, Technology, and the Fate of the Nation. Atlantic Monthly Press. ISBN 978-0-8021-9438-1.

- Hicks, Norman (2011). The Challenge of Economic Development: A Survey of Issues and Constraints Facing Developing Countries. Bloomington,IN: AuthorHouse. p. 59. ISBN 978-1-4567-6633-7.

- American Foundations: An Investigative History; Dowie, Mark; 13 April 2001; MIT Press; Massachusetts; (retrieved from Goodreads online); ISBN 0262041898; accessed March 2014.

- Ponting, Clive (2007). A New Green History of the World: The Environment and the Collapse of Great Civilizations. New York: Penguin Books. p. 244. ISBN 978-0-14-303898-6.

- Morgan, P. (2007). Toilets That Make Compost - Low-cost, sanitary toilets that produce valuable compost for crops in an African context. Stockholm Environment Institute, ISBN 978-9-197-60222-8

- Brown, AD (2003). Feed or feedback: agriculture, population dynamics and the state the planet. International Books. Utrecht, The Netherlands. ISBN 90 5727 048X

- Lüthi, C.; et al. (2011). Sustainable sanitation in cities : a framework for action. Rijswijk: Papiroz Publ. House. p. 38. ISBN 978-90-814088-4-4. Retrieved 19 November 2015.

- Eds.; Simpson-Hébert, co-authors: Uno Winblad, Mayling (2004). Ecological sanitation (2., rev. and enlarged ed.). Stockholm: Stockholm Environment Institute. p. iii. ISBN 9188714985.

- Rosemarin, Arno; McConville, Jennifer; Flores, Amparo; Qiang, Zhu (2012). The challenges of urban ecological sanitation : lessons from the Erdos eco-town project. Practical Action Publishers. p. 116. ISBN 1853397687.

- Mund, Jan-Peter. "Capacities for Megacities coping with water scarcity" (PDF). UN-Water Decade Programme on Capacity Development. Retrieved 2014-02-17.

- Pradhan, Surendra K. "Human Urine and Wood Ash as Plant Nutrients for Red Beet (Beta vulgaris) Cultivation: Impacts on Yield Quality". Journal of Agricultural and Food Chemistry. Retrieved 10 February 2014.

5

Fertilizers and Pesticides in Agricultural Productivity

Fertilizers are materials that aid in the growth of plants. Fertilizers can be classified in several ways, such as single nutrient fertilizers, multinutrient fertilizers and micronutrients. Some of the features of agricultural productivity are pesticide, herbicide, liming, animal feed and irrigation. The chapter on fertilizers and pesticides offers an insightful focus, keeping in mind the topic.

Fertilizer

A large, modern fertilizer spreader

A fertilizer (American English) or fertiliser (British English) is any material of natural or synthetic origin (other than liming materials) that is applied to soils or to plant tissues (usually leaves) to supply one or more plant nutrients essential to the growth of plants.

Mechanism

Six tomato plants grown with and without nitrate fertilizer on nutrient-poor sand/clay soil. One of the plants in the nutrient-poor soil has died.

Fertilizers enhance the growth of plants. This goal is met in two ways, the traditional one being additives that provide nutrients. The second mode by some fertilizers act is to enhance the effectiveness of the soil by modifying its water retention and aeration. This article, like many on fertilizers, emphasises the nutritional aspect. Fertilizers typically provide, in varying proportions:

- three main macronutrients:

    o   Nitrogen (N): leaf growth;

    o   Phosphorus (P): Development of roots, flowers, seeds, fruit;

    o   Potassium (K): Strong stem growth, movement of water in plants, promotion of flowering and fruiting;

- three secondary macronutrients: calcium (Ca), magnesium (Mg), and sulphur (S);

- micronutrients: copper (Cu), iron (Fe), manganese (Mn), molybdenum (Mo), zinc (Zn), boron (B), and of occasional significance there are silicon (Si), cobalt (Co), and vanadium (V) plus rare mineral catalysts.

A Lite-Trac Agri-Spread lime and fertilizer spreader at an agricultural show

The nutrients required for healthy plant life are classified according to the elements, but the elements are not used as fertilizers. Instead compounds containing these elements are the basis of fertilisers. The macronutrients are consumed in larger quantities and are present in plant tissue in quantities from 0.15% to 6.0% on a dry matter (DM) (0% moisture) basis. Plants are made up of four main elements: hydrogen, oxygen, carbon, and nitrogen. Carbon, hydrogen and oxygen are widely available as water and carbon dioxide. Although nitrogen makes up most of the atmosphere, it is in a form that is unavailable to plants. Nitrogen is the most important fertilizer since nitrogen is present in proteins, DNA and other components (e.g., chlorophyll). To be nutritious to plants, nitrogen must be made available in a "fixed" form. Only some bacteria and their host plants (notably legumes) can fix atmospheric nitrogen (N_2) by converting it to ammonia. Phosphate is required for the production of DNA and ATP, the main energy carrier in cells, as well as certain lipids.

Micronutrients are consumed in smaller quantities and are present in plant tissue on the order of parts-per-million (ppm), ranging from 0.15 to 400 ppm DM, or less than 0.04% DM. These elements are often present at the active sites of enzymes that carry out the plant's metabolism. Because these elements enable catalysts (enzymes) their impact far exceeds their weight percentage.

Classification

Fertilizers are classified in several ways. They are classified according to whether they provide a single nutrient (say, N, P, or K), in which case they are classified as "straight fertilizers." "Multinutrient fertilizers" (or "complex fertilizers") provide two or more nutrients, for example N and P. Fertilizers are also sometimes classified as inorganic versus organic. Inorganic fertilizers exclude carbon-containing materials except ureas. Organic fertilizers are usually (recycled) plant- or animal-derived matter. Inorganic are sometimes called synthetic fertilizers since various chemical treatments are required for their manufacture.

Single Nutrient ("Straight") Fertilizers

The main nitrogen-based straight fertilizer is ammonia or its solutions. Ammonium nitrate (NH_4NO_3) is also widely used. About 15M tons were produced in 1981. Urea is another popular source of nitrogen, having the advantage that it is a solid and non-explosive, unlike ammonia and ammonium nitrate, respectively. A few percent of the nitrogen fertilizer market (4% in 2007) has been met by calcium ammonium nitrate ($Ca(NO_3)_2 \cdot NH_4NO_3 \cdot 10H_2O$).

The main straight phosphate fertilizers are the superphosphates. "Single superphosphate" (SSP) consists of 14–18% P_2O_5, again in the form of $Ca(H_2PO_4)_2$, but also phosphogypsum ($CaSO_4 \cdot 2 H_2O$). Triple superphosphate (TSP) typically consists of 44-48% of P_2O_5 and no gypsum. A mixture of single superphosphate and triple superphosphate is called double superphosphate. More than 90% of a typical superphosphate fertilizer is water-soluble.

Multinutrient Fertilizers

These fertilizers are the most common. They consist of two or more nutrient components.

Binary (NP, NK, PK) Fertilizers

Major two-component fertilizers provide both nitrogen and phosphorus to the plants. These are called NP fertilizers. The main NP fertilizers are monoammonium phosphate (MAP) and diammonium phosphate (DAP). The active ingredient in MAP is $NH_4H_2PO_4$. The active ingredient in DAP is $(NH_4)_2HPO_4$. About 85% of MAP and DAP fertilizers are soluble in water.

NPK Fertilizers

NPK fertilizers are three-component fertilizers providing nitrogen, phosphorus, and potassium.

NPK rating is a rating system describing the amount of nitrogen, phosphorus, and potassium in a fertilizer. NPK ratings consist of three numbers separated by dashes (e.g., 10-10-10 or 16-4-8) describing the chemical content of fertilizers. The first number represents the percentage of nitrogen in the product; the second number, P_2O_5; the third, K_2O. Fertilizers do not actually contain P_2O_5 or K_2O, but the system is a conventional shorthand for the amount of the phosphorus (P) or potassium (K) in a fertilizer. A 50-pound (23 kg) bag of fertilizer labeled 16-4-8 contains 8 lb (3.6 kg) of nitrogen (16% of the 50 pounds), an amount of phosphorus equivalent to that in 2 pounds of P_2O_5 (4% of 50 pounds), and 4 pounds of K_2O (8% of 50 pounds). Most fertilizers are labeled according to this N-P-K convention, although Australian convention, following an N-P-K-S system, adds a fourth number for sulfur.

Micronutrients

The main micronutrients are molybdenum, zinc, and copper. These elements are provided as water-soluble salts. Iron presents special problems because it converts to insoluble (bio-unavailable) compounds at moderate soil pH and phosphate concentrations. For this reason, iron is often administered as a chelate complex, e.g., the EDTA derivative. The micronutrient needs depend on the plant. For example, sugar beets appear to require boron, and legumes require cobalt.

Production

Nitrogen Fertilizers

Top users of nitrogen-based fertilizer		
Country	Total N use (Mt pa)	Amt. used for feed/ pasture (Mt pa)
China	18.7	3.0
India	11.9	N/A
U.S.	9.1	4.7
France	2.5	1.3
Germany	2.0	1.2
Brazil	1.7	0.7
Canada	1.6	0.9
Turkey	1.5	0.3
UK	1.3	0.9
Mexico	1.3	0.3
Spain	1.2	0.5
Argentina	0.4	0.1

Nitrogen fertilizers are made from ammonia (NH_3), which is sometimes injected into the ground directly. The ammonia is produced by the Haber-Bosch process. In this energy-intensive process, natural gas (CH_4) supplies the hydrogen, and the nitrogen (N_2) is derived from the air. This ammonia is used as a feedstock for all other nitrogen fertilizers, such as anhydrous ammonium nitrate (NH_4NO_3) and urea ($CO(NH_2)_2$).

Deposits of sodium nitrate ($NaNO_3$) (Chilean saltpeter) are also found in the Atacama desert in Chile and was one of the original (1830) nitrogen-rich fertilizers used. It is still mined for fertilizer.

There has been technical work investigating on-site (on-farm) synthesis of nitrate fertilizer using solar photovoltaic power, which would enable farmers more control in soil fertility, while using far less surface area than conventional organic farming for nitrogen fertilizer.

Phosphate Fertilizers

All phosphate fertilizers are obtained by extraction from minerals containing the anion $PO_4{}^{3-}$. In rare cases, fields are treated with the crushed mineral, but most often more soluble salts are produced by chemical treatment of phosphate minerals. The most popular phosphate-containing minerals are referred to collectively as phosphate rock. The main minerals are fluorapatite $Ca_5(PO_4)_3F$ (CFA) and hydroxyapatite $Ca_5(PO_4)_3OH$. These minerals are converted to water-soluble phosphate salts by treatment with sulfuric or phosphoric acids. The large production of sulfuric acid as an industrial chemical is primarily due to its use as cheap acid in processing phosphate rock into phosphate fertilizer. The global primary uses for both sulfur and phosphorus compounds relate to this basic process.

In the nitrophosphate process or Odda process (invented in 1927), phosphate rock with up to a 20% phosphorus (P) content is dissolved with nitric acid (HNO_3) to produce a mixture of phosphoric acid (H_3PO_4) and calcium nitrate ($Ca(NO_3)_2$). This mixture can be combined with a potassium fertilizer to produce a *compound fertilizer* with the three macronutrients N, P and K in easily dissolved form.

Potassium Fertilizers

Potash is a mixture of potassium minerals used to make potassium (chemical symbol: K) fertilizers. Potash is soluble in water, so the main effort in producing this nutrient from the ore involves some purification steps; e.g., to remove sodium chloride (NaCl) (common salt). Sometimes potash is referred to as K_2O, as a matter of convenience to those describing the potassium content. In fact potash fertilizers are usually potassium chloride, potassium sulfate, potassium carbonate, or potassium nitrate.

Compound Fertilizers

Compound fertilizers, which contain N, P, and K, can often be produced by mixing straight fertilizers. In some cases, chemical reactions occur between the two or more components. For example, monoammonium and diammonium phosphates, which provide plants with both N and P, are produced by neutralizing phosphoric acid (from phosphate rock) and ammonia :

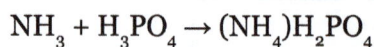

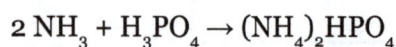

$$NH_3 + H_3PO_4 \rightarrow (NH_4)H_2PO_4$$

$$2\,NH_3 + H_3PO_4 \rightarrow (NH_4)_2HPO_4$$

Organic Fertilizers

Compost bin for small-scale production of organic fertilizer

A large commercial compost operation

The main "organic fertilizers" are peat, animal wastes, plant wastes from agriculture, and treated sewage sludge (biosolids). In terms of volume, peat is the most widely used organic fertilizer. This immature form of coal confers no nutritional value to the plants, but improves the soil by aeration and absorbing water. Animal sources include the products of the slaughter of animals. Bloodmeal, bone meal, hides, hoofs, and horns are typical components. Organic fertilizer usually contain fewer nutrients, but offer other advantages as well as being appealing to those who are trying to practice "environmentally friendly" farming.

Other Elements: Calcium, Magnesium, and Sulfur

Calcium is supplied as superphosphate or calcium ammonium nitrate solutions.

Application

Fertilizers are commonly used for growing all crops, with application rates depending on the soil fertility, usually as measured by a soil test and according to the particular crop. Legumes, for example, fix nitrogen from the atmosphere and generally do not require nitrogen fertilizer.

Liquid Vs Solid

Fertilizers are applied to crops both as solids and as liquid. About 90% of fertilizers are applied as solids. Solid fertilizer is typically granulated or powdered. Often solids are available as prills, a solid globule. Liquid fertilizers comprise anhydrous ammonia, aqueous solutions of ammonia, aqueous solutions of ammonium nitrate or urea. These concentrated products may be diluted with water to form a concentrated liquid fertilizer (e.g., UAN). Advantages of liquid fertilizer are its more rapid effect and easier coverage. The addition of fertilizer to irrigation water is called "fertigation".

Slow- and Controlled-release Fertilizers

Slow- and controlled-release involve only 0.15% (562,000 tons) of the fertilizer market (1995). Their utility stems from the fact that fertilizers are subject to antagonistic processes. In addition to their providing the nutrition to plants, excess fertilizers can be poisonous to the same plant. Competitive with the uptake by plants is the degradation or loss of the fertilizer. Microbes degrade many fertilizers, e.g., by immobilization or oxidation. Furthermore, fertilizers are lost by evaporation or leaching. Most slow-release fertilizers are derivatives of urea, a straight fertilizer providing nitrogen. Isobutylidenediurea ("IBDU") and urea-formaldehyde slowly convert in the soil to free urea, which is rapidly uptaken by plants. IBDU is a single compound with the formula $(CH_3)_2CH-CH(NHC(O)NH_2)_2$ whereas the urea-formaldehydes consist of mixtures of the approximate formula $(HOCH_2NHC(O)NH)_nCH_2$.

Besides being more efficient in the utilization of the applied nutrients, slow-release technologies also reduce the impact on the environment and the contamination of the subsurface water. Slow-release fertilizers (various forms including fertilizer spikes, tabs, etc.) which reduce the problem of "burning" the plants due to excess nitrogen. Polymer coating of fertilizer ingredients gives tablets and spikes a 'true time-release' or 'staged nutrient release' (SNR) of fertilizer nutrients.

Controlled release fertilizers are traditional fertilizers encapsulated in a shell that degrades at a specified rate. Sulfur is a typical encapsulation material. Other coated products use thermoplastics (and sometimes ethylene-vinyl acetate and surfactants, etc.) to produce diffusion-controlled release of urea or other fertilizers. "Reactive Layer Coating" can produce thinner, hence cheaper, membrane coatings by applying reactive monomers simultaneously to the soluble particles. "Multicote" is a process applying layers of low-cost fatty acid salts with a paraffin topcoat.

Foliar Application

Foliar fertilizers are applied directly to leaves. The method is almost invariably used to apply water-soluble straight nitrogen fertilizers and used especially for high value crops such as fruits.

Pesticide

A crop-duster spraying pesticide on a field

A Lite-Trac four-wheeled self-propelled crop sprayer spraying pesticide on a field

Pesticides are substances meant for attracting, seducing, and then destroying any pest. They are a class of biocide. The most common use of pesticides is as plant protection products (also known as crop protection products), which in general protect plants from damaging influences such as weeds, fungi, or insects. This use of pesticides is so common that the term *pesticide* is often treated as synonymous with *plant protection product*, although it is in fact a broader term, as pesticides are also used for non-agricultural purposes. The term pesticide includes all of the following: herbicide, insecticide, insect growth regulator, nematicide, termiticide, molluscicide, piscicide, avicide, rodenticide, predacide, bactericide, insect repellent, animal repellent, antimicrobial, fungicide, disinfectant (antimicrobial), and sanitizer.

In general, a pesticide is a chemical or biological agent (such as a virus, bacterium, antimicrobial, or disinfectant) that deters, incapacitates, kills, or otherwise discourages pests. Target pests can include insects, plant pathogens, weeds, mollusks, birds, mammals, fish, nematodes (roundworms), and microbes that destroy property, cause nuisance, or spread disease, or are disease vectors. Although pesticides have benefits, some also have drawbacks, such as potential toxicity to humans and other species. According to the Stockholm Convention on Persistent Organic Pollutants, 9 of the 12 most dangerous and persistent organic chemicals are organochlorine pesticides.

Definition

Type of pesticide	Target pest group
Herbicides	Plant
Algicides or Algaecides	Algae
Avicides	Birds
Bactericides	Bacteria
Fungicides	Fungi and Oomycetes
Insecticides	Insects
Miticides or Acaricides	Mites
Molluscicides	Snails
Nematicides	Nematodes
Rodenticides	Rodents
Virucides	Viruses

The Food and Agriculture Organization (FAO) has defined *pesticide* as:

any substance or mixture of substances intended for preventing, destroying, or controlling any pest, including vectors of human or animal disease, unwanted species of plants or animals, causing harm during or otherwise interfering with the production, processing, storage, transport, or marketing of food, agricultural commodities, wood and wood products or animal feedstuffs, or substances that may be administered to animals for the control of insects, arachnids, or other pests in or on their bodies. The term includes substances intended for use as a plant growth regulator, defoliant, desiccant, or agent for thinning fruit or preventing the premature fall of fruit. Also used as substances applied to crops either before or after harvest to protect the commodity from deterioration during storage and transport.

Pesticides can be classified by target organism (e.g., herbicides, insecticides, fungicides, rodenticides, and pediculicides), chemical structure (e.g., organic, inorganic, synthetic, or biological (biopesticide), although the distinction can sometimes blur), and physical state (e.g. gaseous (fumigant)). Biopesticides include microbial pesticides and biochemical pesticides. Plant-derived pesticides, or "botanicals", have been developing quickly. These include the pyrethroids, rotenoids, nicotinoids, and a fourth group that includes strychnine and scilliroside.

Many pesticides can be grouped into chemical families. Prominent insecticide families include organochlorines, organophosphates, and carbamates. Organochlorine hydrocarbons (e.g., DDT) could be separated into dichlorodiphenylethanes, cyclodiene compounds, and other related compounds. They operate by disrupting the sodium/potassium balance of the nerve fiber, forcing the nerve to transmit continuously. Their toxicities vary greatly, but they have been phased out because of their persistence and potential to bioaccumulate. Organophosphate and carbamates largely replaced organochlorines. Both operate through inhibiting the enzyme acetylcholinesterase, allowing acetylcholine to transfer nerve impulses indefinitely and causing a variety of symptoms such as weakness or paralysis. Organophosphates are quite toxic to vertebrates, and have in some cases been replaced by less toxic carbamates. Thiocarbamate and dithiocarbamates are subclasses of carbamates. Prominent families of herbicides include phenoxy and benzoic acid herbicides (e.g. 2,4-D), triazines (e.g., atrazine), ureas (e.g., diuron), and Chloroacetanilides (e.g., alachlor). Phenoxy compounds tend to selectively kill broad-leaf weeds rather than grasses. The phenoxy and benzoic acid herbicides function similar to plant growth hormones, and grow cells without normal cell division, crushing the plant's nutrient transport system. Triazines interfere with photosynthesis. Many commonly used pesticides are not included in these families, including glyphosate.

Pesticides can be classified based upon their biological mechanism function or application method. Most pesticides work by poisoning pests. A systemic pesticide moves inside a plant following absorption by the plant. With insecticides and most fungicides, this movement is usually upward (through the xylem) and outward. Increased efficiency may be a result. Systemic insecticides, which poison pollen and nectar in the flowers, may kill bees and other needed pollinators.

In 2009, the development of a new class of fungicides called paldoxins was announced. These work by taking advantage of natural defense chemicals released by plants called phytoalexins, which fungi then detoxify using enzymes. The paldoxins inhibit the fungi's detoxification enzymes. They are believed to be safer and greener.

Uses

Pesticides are used to control organisms that are considered to be harmful. For example, they are used to kill mosquitoes that can transmit potentially deadly diseases like West Nile virus, yellow fever, and malaria. They can also kill bees, wasps or ants that can cause allergic reactions. Insecticides can protect animals from illnesses that can be caused by parasites such as fleas. Pesticides can prevent sickness in humans that could be caused by moldy food or diseased produce. Herbicides can be used to clear roadside weeds, trees and brush. They can also kill invasive weeds that may cause environmental damage. Herbicides are commonly applied in ponds and lakes to control algae and plants such as water grasses that can interfere with activities like swimming and fishing and cause the water to look or smell unpleasant. Uncontrolled pests such as termites and mold can damage structures such as houses. Pesticides are used in grocery stores and food storage facilities to manage rodents and insects that infest food such as grain. Each use of a pesticide carries some associated risk. Proper pesticide use decreases these associated risks to a level deemed acceptable by pesticide regulatory agencies such as the United States Environmental Protection Agency (EPA) and the Pest Management Regulatory Agency (PMRA) of Canada.

DDT, sprayed on the walls of houses, is an organochlorine that has been used to fight malaria since the 1950s. Recent policy statements by the World Health Organization have given stronger support to this approach. However, DDT and other organochlorine pesticides have been banned in most countries worldwide because of their persistence in the environment and human toxicity. DDT use is not always effective, as resistance to DDT was identified in Africa as early as 1955, and by 1972 nineteen species of mosquito worldwide were resistant to DDT.

Amount Used

In 2006 and 2007, the world used approximately 2.4 megatonnes (5.3×10^9 lb) of pesticides, with herbicides constituting the biggest part of the world pesticide use at 40%, followed by insecticides (17%) and fungicides (10%). In 2006 and 2007 the U.S. used approximately 0.5 megatonnes (1.1×10^9 lb) of pesticides, accounting for 22% of the world total, including 857 million pounds (389 kt) of conventional pesticides, which are used in the agricultural sector (80% of conventional pesticide use) as well as the industrial, commercial, governmental and home & garden sectors. Pesticides are also found in majority of U.S. households with 78 million out of the 105.5 million households indicating that they use some form of pesticide. As of 2007, there were more than 1,055 active ingredients registered as pesticides, which yield over 20,000 pesticide products that are marketed in the United States.

The US used some 1 kg (2.2 pounds) per hectare of arable land compared with: 4.7 kg in China, 1.3 kg in the UK, 0.1 kg in Cameroon, 5.9 kg in Japan and 2.5 kg in Italy. Insecticide use in the US has declined by more than half since 1980, (.6%/yr) mostly due to the near phase-out of organophosphates. In corn fields, the decline was even steeper, due to the switchover to transgenic Bt corn.

For the global market of crop protection products, market analysts forecast revenues of over 52 billion US$ in 2019.

Benefits

Pesticides can save farmers' money by preventing crop losses to insects and other pests; in the U.S., farmers get an estimated fourfold return on money they spend on pesticides. One study found that not using pesticides reduced crop yields by about 10%. Another study, conducted in 1999, found that a ban on pesticides in the United States may result in a rise of food prices, loss of jobs, and an increase in world hunger.

There are two levels of benefits for pesticide use, primary and secondary. Primary benefits are direct gains from the use of pesticides and secondary benefits are effects that are more long-term.

Primary Benefits

1. Controlling pests and plant disease vectors

    o   Improved crop/livestock yields

    o   Improved crop/livestock quality

    o   Invasive species controlled

2. Controlling human/livestock disease vectors and nuisance organisms

    o   Human lives saved and suffering reduced

    o   Animal lives saved and suffering reduced

    o   Diseases contained geographically

3. Controlling organisms that harm other human activities and structures

    o   Drivers view unobstructed

    o   Tree/brush/leaf hazards prevented

    o   Wooden structures protected

Monetary

Every dollar ($1) that is spent on pesticides for crops yields four dollars ($4) in crops saved. This means based that, on the amount of money spent per year on pesticides, $10 billion, there is an additional $40 billion savings in crop that would be lost due to damage by insects and weeds. In general, farmers benefit from having an increase in crop yield and from being able to grow a variety of crops throughout the year. Consumers of agricultural products also benefit from being able to afford the vast quantities of produce available year-round. The general public also benefits from the use of pesticides for the control of insect-borne diseases and illnesses, such as malaria. The use of pesticides creates a large job market within the agrichemical sector.

Costs

On the cost side of pesticide use there can be costs to the environment, costs to human health, as well as costs of the development and research of new pesticides.

Health Effects

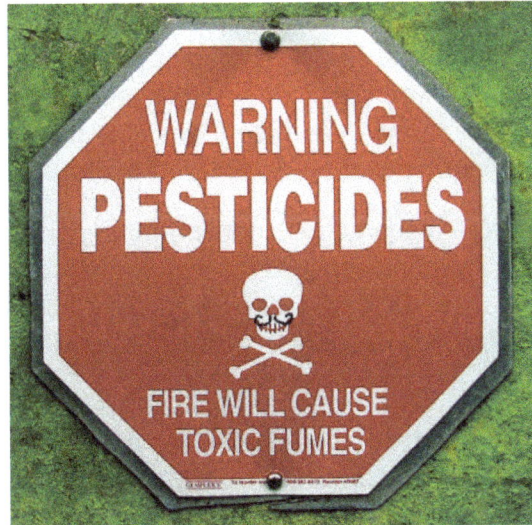

A sign warning about potential pesticide exposure.

Pesticides may cause acute and delayed health effects in people who are exposed. Pesticide exposure can cause a variety of adverse health effects, ranging from simple irritation of the skin and eyes to more severe effects such as affecting the nervous system, mimicking hormones causing reproductive problems, and also causing cancer. A 2007 systematic review found that "most studies on non-Hodgkin lymphoma and leukemia showed positive associations with pesticide exposure" and thus concluded that cosmetic use of pesticides should be decreased. There is substantial evidence of associations between organophosphate insecticide exposures and neurobehavioral alterations. Limited evidence also exists for other negative outcomes from pesticide exposure including neurological, birth defects, and fetal death.

The American Academy of Pediatrics recommends limiting exposure of children to pesticides and using safer alternatives:

The World Health Organization and the UN Environment Programme estimate that each year, 3 million workers in agriculture in the developing world experience severe poisoning from pesticides, about 18,000 of whom die. Owing to inadequate regulation and safety precautions, 99% of pesticide related deaths occur in developing countries that account for only 25% of pesticide usage. According to one study, as many as 25 million workers in developing countries may suffer mild pesticide poisoning yearly. There are several careers aside from agriculture that may also put individuals at risk of health effects from pesticide exposure including pet groomers, groundskeepers, and fumigators.

One study found pesticide self-poisoning the method of choice in one third of suicides worldwide, and recommended, among other things, more restrictions on the types of pesticides that are most harmful to humans.

A 2014 epidemiological review found associations between autism and exposure to certain pesticides, but noted that the available evidence was insufficient to conclude that the relationship was causal.

Environmental Effect

Pesticide use raises a number of environmental concerns. Over 98% of sprayed insecticides and 95% of herbicides reach a destination other than their target species, including non-target species, air, water and soil. Pesticide drift occurs when pesticides suspended in the air as particles are carried by wind to other areas, potentially contaminating them. Pesticides are one of the causes of water pollution, and some pesticides are persistent organic pollutants and contribute to soil contamination.

In addition, pesticide use reduces biodiversity, contributes to pollinator decline, destroys habitat (especially for birds), and threatens endangered species. Pests can develop a resistance to the pesticide (pesticide resistance), necessitating a new pesticide. Alternatively a greater dose of the pesticide can be used to counteract the resistance, although this will cause a worsening of the ambient pollution problem.

Since chlorinated hydrocarbon pesticides dissolve in fats and are not excreted, organisms tend to retain them almost indefinitely. Biological magnification is the process whereby these chlorinated hydrocarbons (pesticides) are more concentrated at each level of the food chain. Among marine animals, pesticide concentrations are higher in carnivorous fishes, and even more so in the fish-eating birds and mammals at the top of the ecological pyramid. Global distillation is the process whereby pesticides are transported from warmer to colder regions of the Earth, in particular the Poles and mountain tops. Pesticides that evaporate into the atmosphere at relatively high temperature can be carried considerable distances (thousands of kilometers) by the wind to an area of lower temperature, where they condense and are carried back to the ground in rain or snow.

In order to reduce negative impacts, it is desirable that pesticides be degradable or at least quickly deactivated in the environment. Such loss of activity or toxicity of pesticides is due to both innate chemical properties of the compounds and environmental processes or conditions. For example, the presence of halogens within a chemical structure often slows down degradation in an aerobic environment. Adsorption to soil may retard pesticide movement, but also may reduce bioavailability to microbial degraders.

Economics

Harm	Annual US cost
Public health	$1.1 billion
Pesticide resistance in pest	$1.5 billion
Crop losses caused by pesticides	$1.4 billion
Bird losses due to pesticides	$2.2 billion
Groundwater contamination	$2.0 billion
Other costs	$1.4 billion
Total costs	**$9.6 billion**

Human health and environmental cost from pesticides in the United States is estimated at $9.6 billion offset by about $40 billion in increased agricultural production:

Additional costs include the registration process and the cost of purchasing pesticides. The registration process can take several years to complete (there are 70 different types of field test) and can cost $50–70 million for a single pesticide. Annually the United States spends $10 billion on pesticides.

Alternatives

Alternatives to pesticides are available and include methods of cultivation, use of biological pest controls (such as pheromones and microbial pesticides), genetic engineering, and methods of interfering with insect breeding. Application of composted yard waste has also been used as a way of controlling pests. These methods are becoming increasingly popular and often are safer than traditional chemical pesticides. In addition, EPA is registering reduced-risk conventional pesticides in increasing numbers.

Cultivation practices include polyculture (growing multiple types of plants), crop rotation, planting crops in areas where the pests that damage them do not live, timing planting according to when pests will be least problematic, and use of trap crops that attract pests away from the real crop. In the U.S., farmers have had success controlling insects by spraying with hot water at a cost that is about the same as pesticide spraying.

Release of other organisms that fight the pest is another example of an alternative to pesticide use. These organisms can include natural predators or parasites of the pests. Biological pesticides based on entomopathogenic fungi, bacteria and viruses cause disease in the pest species can also be used.

Interfering with insects' reproduction can be accomplished by sterilizing males of the target species and releasing them, so that they mate with females but do not produce offspring. This technique was first used on the screwworm fly in 1958 and has since been used with the medfly, the tsetse fly, and the gypsy moth. However, this can be a costly, time consuming approach that only works on some types of insects.

Agroecology emphasize nutrient recycling, use of locally available and renewable resources, adaptation to local conditions, utilization of microenvironments, reliance on indigenous knowledge and yield maximization while maintaining soil productivity. Agroecology also emphasizes empowering people and local communities to contribute to development, and encouraging "multi-directional" communications rather than the conventional "top-down" method.

Push Pull Strategy

The term "push-pull" was established in 1987 as an approach for integrated pest management (IPM). This strategy uses a mixture of behavior-modifying stimuli to manipulate the distribution and abundance of insects. "Push" means the insects are repelled or deterred away from whatever resource that is being protected. "Pull" means that certain stimuli (semiochemical stimuli, pheromones, food additives, visual stimuli, genetically altered plants, etc.) are used to attract pests to trap crops where they will be killed. There are numerous different components involved in order to implement a Push-Pull Strategy in IPM.

Many case studies testing the effectiveness of the push-pull approach have been done across the world. The most successful push-pull strategy was developed in Africa for subsistence farming. Another successful case study was performed on the control of *Helicoverpa* in cotton crops in Australia. In Europe, the Middle East, and the United States, push-pull strategies were successfully used in the controlling of *Sitona lineatus* in bean fields.

Some advantages of using the push-pull method are less use of chemical or biological materials and better protection against insect habituation to this control method. Some disadvantages of the push-pull strategy is that if there is a lack of appropriate knowledge of behavioral and chemical ecology of the host-pest interactions then this method becomes unreliable. Furthermore, because the push-pull method is not a very popular method of IPM operational and registration costs are higher.

Effectiveness

Some evidence shows that alternatives to pesticides can be equally effective as the use of chemicals. For example, Sweden has halved its use of pesticides with hardly any reduction in crops.In Indonesia, farmers have reduced pesticide use on rice fields by 65% and experienced a 15% crop increase.A study of Maize fields in northern Florida found that the application of composted yard waste with high carbon to nitrogen ratio to agricultural fields was highly effective at reducing the population of plant-parasitic nematodes and increasing crop yield, with yield increases ranging from 10% to 212%; the observed effects were long-term, often not appearing until the third season of the study.

However, pesticide resistance is increasing. In the 1940s, U.S. farmers lost only 7% of their crops to pests. Since the 1980s, loss has increased to 13%, even though more pesticides are being used.Between 500 and 1,000 insect and weed species have developed pesticide resistance since 1945.

Types

Pesticides are often referred to according to the type of pest they control. Pesticides can also be considered as either biodegradable pesticides, which will be broken down by microbes and other living beings into harmless compounds, or persistent pesticides, which may take months or years before they are broken down: it was the persistence of DDT, for example, which led to its accumulation in the food chain and its killing of birds of prey at the top of the food chain. Another way to think about pesticides is to consider those that are chemical pesticides or are derived from a common source or production method.

Some examples of chemically-related pesticides are:

Organophosphate Pesticides

Organophosphates affect the nervous system by disrupting, acetylcholinesterase activity, the enzyme that regulates acetylcholine, a neurotransmitter. Most organophosphates are insecticides. They were developed during the early 19th century, but their effects on insects, which are similar to their effects on humans, were discovered in 1932. Some are very poisonous. However, they usually are not persistent in the environment.

Carbamate Pesticides

Carbamate pesticides affect the nervous system by disrupting an enzyme that regulates acetylcholine, a neurotransmitter. The enzyme effects are usually reversible. There are several subgroups within the carbamates.

Organochlorine Insecticides

They were commonly used in the past, but many have been removed from the market due to their health and environmental effects and their persistence (e.g., DDT, chlordane, and toxaphene).

Pyrethroid Pesticides

They were developed as a synthetic version of the naturally occurring pesticide pyrethrin, which is found in chrysanthemums. They have been modified to increase their stability in the environment. Some synthetic pyrethroids are toxic to the nervous system.

Sulfonylurea Herbicides

The following sulfonylureas have been commercialized for weed control: amidosulfuron, azimsulfuron, bensulfuron-methyl, chlorimuron-ethyl, ethoxysulfuron, flazasulfuron, flupyrsulfuron-methyl-sodium, halosulfuron-methyl, imazosulfuron, nicosulfuron, oxasulfuron, primisulfuron-methyl, pyrazosulfuron-ethyl, rimsulfuron, sulfometuron-methyl Sulfosulfuron, terbacil, bispyribac-sodium, cyclosulfamuron, and pyrithiobac-sodium. Nicosulfuron, triflusulfuron methyl, and chlorsulfuron are broad-spectrum herbicides that kill plants by inhibiting the enzyme acetolactate synthase. In the 1960s, more than 1 kg/ha (0.89 lb/acre) crop protection chemical was typically applied, while sulfonylureates allow as little as 1% as much material to achieve the same effect.

Biopesticides

Biopesticides are certain types of pesticides derived from such natural materials as animals, plants, bacteria, and certain minerals. For example, canola oil and baking soda have pesticidal applications and are considered biopesticides. Biopesticides fall into three major classes:

- Microbial pesticides which consist of bacteria, entomopathogenic fungi or viruses (and sometimes includes the metabolites that bacteria or fungi produce). Entomopathogenic nematodes are also often classed as microbial pesticides, even though they are multi-cellular.

- Biochemical pesticides or herbal pesticides are naturally occurring substances that control (or monitor in the case of pheromones) pests and microbial diseases.

- Plant-incorporated protectants (PIPs) have genetic material from other species incorporated into their genetic material (*i.e.* GM crops). Their use is controversial, especially in many European countries.

Classified By Type of Pest

Pesticides that are related to the type of pests are:

Type	Action
Algicides	Control algae in lakes, canals, swimming pools, water tanks, and other sites
Antifouling agents	Kill or repel organisms that attach to underwater surfaces, such as boat bottoms
Antimicrobials	Kill microorganisms (such as bacteria and viruses)
Attractants	Attract pests (for example, to lure an insect or rodent to a trap). (However, food is not considered a pesticide when used as an attractant.)
Biopesticides	Biopesticides are certain types of pesticides derived from such natural materials as animals, plants, bacteria, and certain minerals
Biocides	Kill microorganisms
Disinfectants and sanitizers	Kill or inactivate disease-producing microorganisms on inanimate objects
Fungicides	Kill fungi (including blights, mildews, molds, and rusts)
Fumigants	Produce gas or vapor intended to destroy pests in buildings or soil
Herbicides	Kill weeds and other plants that grow where they are not wanted
Insecticides	Kill insects and other arthropods
Miticides	Kill mites that feed on plants and animals
Microbial pesticides	Microorganisms that kill, inhibit, or out compete pests, including insects or other microorganisms
Molluscicides	Kill snails and slugs
Nematicides	Kill nematodes (microscopic, worm-like organisms that feed on plant roots)
Ovicides	Kill eggs of insects and mites
Pheromones	Biochemicals used to disrupt the mating behavior of insects
Repellents	Repel pests, including insects (such as mosquitoes) and birds
Rodenticides	Control mice and other rodents

Further Types of Pesticides

The term pesticide also include these substances:

Defoliants : Cause leaves or other foliage to drop from a plant, usually to facilitate harvest. Desiccants : Promote drying of living tissues, such as unwanted plant tops. Insect growth regulators : Disrupt the molting, maturity from pupal stage to adult, or other life processes of insects. Plant growth regulators : Substances (excluding fertilizers or other plant nutrients) that alter the expected growth, flowering, or reproduction rate of plants.

Regulation

International

In most countries, pesticides must be approved for sale and use by a government agency.

In Europe, recent EU legislation has been approved banning the use of highly toxic pesticides including those that are carcinogenic, mutagenic or toxic to reproduction, those that are endocrine-disrupting, and those that are persistent, bioaccumulative and toxic (PBT) or very persistent and very bioaccumulative (vPvB).Measures were approved to improve the general safety of pesticides across all EU member states.

Though pesticide regulations differ from country to country, pesticides, and products on which they were used are traded across international borders. To deal with inconsistencies in regulations among countries, delegates to a conference of the United Nations Food and Agriculture Organization adopted an International Code of Conduct on the Distribution and Use of Pesticides in 1985 to create voluntary standards of pesticide regulation for different countries. The Code was updated in 1998 and 2002. The FAO claims that the code has raised awareness about pesticide hazards and decreased the number of countries without restrictions on pesticide use.

Three other efforts to improve regulation of international pesticide trade are the United Nations London Guidelines for the Exchange of Information on Chemicals in International Trade and the United Nations Codex Alimentarius Commission. The former seeks to implement procedures for ensuring that prior informed consent exists between countries buying and selling pesticides, while the latter seeks to create uniform standards for maximum levels of pesticide residues among participating countries. Both initiatives operate on a voluntary basis.

Pesticides safety education and pesticide applicator regulation are designed to protect the public from pesticide misuse, but do not eliminate all misuse. Reducing the use of pesticides and choosing less toxic pesticides may reduce risks placed on society and the environment from pesticide use. Integrated pest management, the use of multiple approaches to control pests, is becoming widespread and has been used with success in countries such as Indonesia, China, Bangladesh, the U.S., Australia, and Mexico. IPM attempts to recognize the more widespread impacts of an action on an ecosystem, so that natural balances are not upset. New pesticides are being developed, including biological and botanical derivatives and alternatives that are thought to reduce health and environmental risks. In addition, applicators are being encouraged to consider alternative controls and adopt methods that reduce the use of chemical pesticides.

Pesticides can be created that are targeted to a specific pest's lifecycle, which can be environmentally more friendly. For example, potato cyst nematodes emerge from their protective cysts in response to a chemical excreted by potatoes; they feed on the potatoes and damage the crop. A similar chemical can be applied to fields early, before the potatoes are planted, causing the nematodes to emerge early and starve in the absence of potatoes.

United States

Preparation for an application of hazardous herbicide in USA.

In the United States, the Environmental Protection Agency (EPA) is responsible for regulating pesticides under the Federal Insecticide, Fungicide, and Rodenticide Act (FIFRA) and the Food Quality Protection Act (FQPA). Studies must be conducted to establish the conditions in which the material is safe to use and the effectiveness against the intended pest(s). The EPA regulates pesticides to ensure that these products do not pose adverse effects to humans or the environment. Pesticides produced before November 1984 continue to be reassessed in order to meet the current scientific and regulatory standards. All registered pesticides are reviewed every 15 years to ensure they meet the proper standards. During the registration process, a label is created. The label contains directions for proper use of the material in addition to safety restrictions. Based on acute toxicity, pesticides are assigned to a Toxicity Class.

Some pesticides are considered too hazardous for sale to the general public and are designated restricted use pesticides. Only certified applicators, who have passed an exam, may purchase or supervise the application of restricted use pesticides. Records of sales and use are required to be maintained and may be audited by government agencies charged with the enforcement of pesticide regulations. These records must be made available to employees and state or territorial environmental regulatory agencies.

The EPA regulates pesticides under two main acts, both of which amended by the Food Quality Protection Act of 1996. In addition to the EPA, the United States Department of Agriculture (USDA) and the United States Food and Drug Administration (FDA) set standards for the level of pesticide residue that is allowed on or in crops. The EPA looks at what the potential human health and environmental effects might be associated with the use of the pesticide.

In addition, the U.S. EPA uses the National Research Council's four-step process for human health risk assessment: (1) Hazard Identification, (2) Dose-Response Assessment, (3) Exposure Assessment, and (4) Risk Characterization.

Recently Kaua'i County (Hawai'i) passed Bill No. 2491 to add an article to Chapter 22 of the county's code relating to pesticides and GMOs. The bill strengthens protections of local communities in Kaua'i where many large pesticide companies test their products.

History

Since before 2000 BC, humans have utilized pesticides to protect their crops. The first known pesticide was elemental sulfur dusting used in ancient Sumer about 4,500 years ago in ancient Mesopotamia. The Rig Veda, which is about 4,000 years old, mentions the use of poisonous plants for pest control. By the 15th century, toxic chemicals such as arsenic, mercury, and lead were being applied to crops to kill pests. In the 17th century, nicotine sulfate was extracted from tobacco leaves for use as an insecticide. The 19th century saw the introduction of two more natural pesticides, pyrethrum, which is derived from chrysanthemums, and rotenone, which is derived from the roots of tropical vegetables. Until the 1950s, arsenic-based pesticides were dominant. Paul Müller discovered that DDT was a very effective insecticide. Organochlorines such as DDT were dominant, but they were replaced in the U.S. by organophosphates and carbamates by 1975. Since then, pyrethrin compounds have become the dominant insecticide. Herbicides became common in the 1960s, led by "triazine and other nitrogen-based compounds, carboxylic acids such as 2,4-dichlorophenoxyacetic acid, and glyphosate".

The first legislation providing federal authority for regulating pesticides was enacted in 1910; however, decades later during the 1940s manufacturers began to produce large amounts of synthetic pesticides and their use became widespread. Some sources consider the 1940s and 1950s to have been the start of the "pesticide era." Although the U.S. Environmental Protection Agency was established in 1970 and amendments to the pesticide law in 1972, pesticide use has increased 50-fold since 1950 and 2.3 million tonnes (2.5 million short tons) of industrial pesticides are now used each year. Seventy-five percent of all pesticides in the world are used in developed countries, but use in developing countries is increasing. A study of USA pesticide use trends through 1997 was published in 2003 by the National Science Foundation's Center for Integrated Pest Management.

In the 1960s, it was discovered that DDT was preventing many fish-eating birds from reproducing, which was a serious threat to biodiversity. Rachel Carson wrote the best-selling book *Silent Spring* about biological magnification. The agricultural use of DDT is now banned under the Stockholm Convention on Persistent Organic Pollutants, but it is still used in some developing nations to prevent malaria and other tropical diseases by spraying on interior walls to kill or repel mosquitoes.

Herbicide

Weeds controlled with herbicide

Herbicide(s), also commonly known as weedkillers, are chemical substances used to control unwanted plants. Selective herbicides control specific weed species, while leaving the desired crop relatively unharmed, while non-selective herbicides (sometimes called "total weedkillers" in commercial products) can be used to clear waste ground, industrial and construction sites, railways and railway embankments as they kill all plant material with which they come into contact. Apart from selective/non-selective, other important distinctions include *persistence* (also known as *residual action*: how long the product stays in place and remains active), *means of uptake* (whether it is absorbed by above-ground foliage only, through the roots, or by other means), and *mechanism of action* (how it works). Historically, products such as common salt and other metal salts were used as herbicides, however these have gradually fallen out of favor and in some countries a number of these are banned due to their persistence in soil, and toxicity and groundwater contamination concerns. Herbicides have also been used in warfare and conflict.

Modern herbicides are often synthetic mimics of natural plant hormones which interfere with growth of the target plants. The term organic herbicide has come to mean herbicides intended for organic farming; these are often less efficient and more costly than synthetic herbicides and are based on natural materials. Some plants also produce their own natural herbicides, such as the genus *Juglans* (walnuts), or the tree of heaven; such action of natural herbicides, and other related chemical interactions, is called allelopathy. Due to herbicide resistance - a major concern in agriculture - a number of products also combine herbicides with different means of action.

In the US in 2007, about 83% of all herbicide usage, determined by weight applied, was in agriculture. In 2007, world pesticide expenditures totaled about $39.4 billion; herbicides were about 40% of those sales and constituted the biggest portion, followed by insecticides, fungicides, and other types. Smaller quantities are used in forestry, pasture systems, and management of areas set aside as wildlife habitat.

History

Prior to the widespread use of chemical herbicides, cultural controls, such as altering soil pH, salinity, or fertility levels, were used to control weeds. Mechanical control (including tillage) was also (and still is) used to control weeds.

First Herbicides

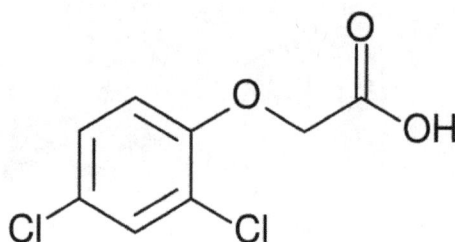

2,4-D, the first chemical herbicide, was discovered during the Second World War.

Although research into chemical herbicides began in the early 20th century, the first major breakthrough was the result of research conducted in both the UK and the US during the Second World War into the potential use of agents as biological weapons. The first modern herbicide, 2,4-D, was first discovered and synthesized by W. G. Templeman at Imperial Chemical Industries. In 1940, he showed that "Growth substances applied appropriately would kill certain broad-leaved weeds in cereals without harming the crops." By 1941, his team succeeded in synthesizing the chemical. In the same year, Pokorny in the US achieved this as well.

Independently, a team under Juda Hirsch Quastel, working at the Rothamsted Experimental Station made the same discovery. Quastel was tasked by the Agricultural Research Council (ARC) to discover methods for improving crop yield. By analyzing soil as a dynamic system, rather than an inert substance, he was able to apply techniques such as perfusion. Quastel was able to quantify the influence of various plant hormones, inhibitors and other chemicals on the activity of microorganisms in the soil and assess their direct impact on plant growth. While the full work of the unit remained secret, certain discoveries were developed for commercial use after the war, including the 2,4-D compound.

When it was commercially released in 1946, it triggered a worldwide revolution in agricultural output and became the first successful selective herbicide. It allowed for greatly enhanced weed control in wheat, maize (corn), rice, and similar cereal grass crops, because it kills dicots (broadleaf plants), but not most monocots (grasses). The low cost of 2,4-D has led to continued usage today, and it remains one of the most commonly used herbicides in the world. Like other acid herbicides, current formulations use either an amine salt (often trimethylamine) or one of many esters of the parent compound. These are easier to handle than the acid.

Further Discoveries

The triazine family of herbicides, which includes atrazine, were introduced in the 1950s; they have the current distinction of being the herbicide family of greatest concern regarding groundwater contamination. Atrazine does not break down readily (within a few weeks) after being applied to soils of above neutral pH. Under alkaline soil conditions, atrazine may be carried into the soil profile as far as the water table by soil water following rainfall causing the aforementioned contamination. Atrazine is thus said to have "carryover", a generally undesirable property for herbicides.

Glyphosate (Roundup) was introduced in 1974 for nonselective weed control. Following the development of glyphosate-resistant crop plants, it is now used very extensively for selective weed control in growing crops. The pairing of the herbicide with the resistant seed contributed to the consolidation of the seed and chemistry industry in the late 1990s.

Many modern chemical herbicides used in agriculture and gardening are specifically formulated to decompose within a short period after application. This is desirable, as it allows crops and plants to be planted afterwards, which could otherwise be affected by the herbicide. However, herbicides with low residual activity (i.e., that decompose quickly) often do not provide season-long weed control and do not ensure that weed roots are killed beneath construction and paving (and cannot emerge destructively in years to come), therefore there remains a role for weedkiller with high levels of persistence in the soil.

Terminology

Herbicides are classified/grouped in various ways e.g. according to the activity, timing of application, method of application, mechanism of action, chemical family. This gives rise to a considerable level of terminology related to herbicides and their use.

Intended Outcome

- Control is the destruction of unwanted weeds, or the damage of them to the point where they are no longer competitive with the crop.

- Suppression is incomplete control still providing some economic benefit, such as reduced competition with the crop.

- Crop safety, for selective herbicides, is the relative absence of damage or stress to the crop. Most selective herbicides cause some visible stress to crop plants.

- Defoliant, similar to herbicides, but designed to remove foliage (leaves) rather than kill the plant.

Selectivity (All Plants or Specific Plants)

- Selective herbicides: They control or suppress certain plants without affecting the growth of other plants species. Selectivity may be due to translocation, differential absorption, physical (morphological) or physiological differences between plant species. 2,4-D, mecoprop, dicamba control many broadleaf weeds but remain ineffective against turfgrasses.

- Non-selective herbicides: These herbicides are not specific in acting against certain plant species and control all plant material with which they come into contact. They are used to clear industrial sites, waste ground, railways and railway embankments. Paraquat, glufosinate, glyphosate are non-selective herbicides.

Timing of Application

- Preplant: Preplant herbicides are nonselective herbicides applied to soil before planting. Some preplant herbicides may be mechanically incorporated into the soil. The objective for incorporation is to prevent dissipation through photodecomposition and/or volatility. The herbicides kill weeds as they grow through the herbicide treated zone. Volatile herbicides have to be incorporated into the soil before planting the pasture. Agricultural crops grown in soil treated with a preplant herbicide include tomatoes, corn, soybeans and strawberries. Soil fumigants like metam-sodium and dazomet are in use as preplant herbicides.

- Preemergence: Preemergence herbicides are applied before the weed seedlings emerge through the soil surface. Herbicides do not prevent weeds from germinating but they kill weeds as they grow through the herbicide treated zone by affecting the cell division in the emerging seedling. Dithopyr and pendimethalin are preemergence herbicides. Weeds that have already emerged before application or activation are not affected by pre-herbicides as their primary growing point escapes the treatment.

- Postemergence: These herbicides are applied after weed seedlings have emerged through the soil surface. They can be foliar or root absorbed, selective or nonselective, contact or systemic. Application of these herbicides is avoided during rain because the problem of being washed off to the soil makes it ineffective. 2,4-D is a selective, systemic, foliar absorbed postemergence herbicide.

Method of Application

- Soil applied: Herbicides applied to the soil are usually taken up by the root or shoot of the emerging seedlings and are used as preplant or preemergence treatment. Several factors influence the effectiveness of soil-applied herbicides. Weeds absorb herbicides by both passive and active mechanism. Herbicide adsorption to soil colloids or organic matter often reduces its amount available for weed absorption. Positioning of herbicide in correct layer of soil is very important, which can be achieved mechanically and by rainfall. Herbicides on the soil surface are subjected to several processes that reduce their availability. Volatility and photolysis are two common processes that reduce the availability of herbicides. Many soil applied herbicides are absorbed through plant shoots while they are still underground leading to their death or injury. EPTC and trifluralin are soil applied herbicides.

- Foliar applied: These are applied to portion of the plant above the ground and are absorbed by exposed tissues. These are generally postemergence herbicides and can either be translocated (systemic) throughout the plant or remain at specific site (contact). External barriers of plants like cuticle, waxes, cell wall etc. affect herbicide absorption and action. Glyphosate, 2,4-D and dicamba are foliar applied herbicide.

Persistence

- Residual activity: A herbicide is described as having low residual activity if it is neutralized within a short time of application (within a few weeks or months) - typically this is due to rainfall, or by reactions in the soil. A herbicide described as having high residual activity will remains potent for a long term in the soil. For some compounds, the residual activity can leave the ground almost permanently barren.

Mechanism of Action

Herbicides are often classified according to their site of action, because as a general rule, herbicides within the same site of action class will produce similar symptoms on susceptible plants. Classification based on site of action of herbicide is comparatively better as herbicide resistance management can be handled more properly and effectively. Classification by mechanism of action (MOA) indicates the first enzyme, protein, or biochemical step affected in the plant following application.

List of Mechanisms found in Modern Herbicides

- ACCase inhibitors compounds kill grasses. Acetyl coenzyme A carboxylase (ACCase) is part of the first step of lipid synthesis. Thus, ACCase inhibitors affect cell membrane production in the meristems of the grass plant. The ACCases of grasses are sensitive to these herbicides, whereas the ACCases of dicot plants are not.

- ALS inhibitors: the acetolactate synthase (ALS) enzyme (also known as acetohydroxyacid synthase, or AHAS) is the first step in the synthesis of the branched-chain amino acids (valine, leucine, and isoleucine). These herbicides slowly starve affected plants of these amino acids, which eventually leads to inhibition of DNA synthesis. They affect grasses and dicots alike. The ALS inhibitor family includes various sulfonylureas (such as Flazasulfuron and Metsulfuron-methyl), imidazolinones, triazolopyrimidines, pyrimidinyl oxybenzoates, and sulfonylamino carbonyl triazolinones. The ALS biological pathway exists only in plants and not animals, thus making the ALS-inhibitors among the safest herbicides.

- EPSPS inhibitors: The enolpyruvylshikimate 3-phosphate synthase enzyme EPSPS is used in the synthesis of the amino acids tryptophan, phenylalanine and tyrosine. They affect grasses and dicots alike. Glyphosate (Roundup) is a systemic EPSPS inhibitor inactivated by soil contact.

- Synthetic auxins inaugurated the era of organic herbicides. They were discovered in the 1940s after a long study of the plant growth regulator auxin. Synthetic auxins mimic this plant hormone. They have several points of action on the cell membrane, and are effective in the control of dicot plants. 2,4-D is a synthetic auxin herbicide.

- Photosystem II inhibitors reduce electron flow from water to NADPH2+ at the photochemical step in photosynthesis. They bind to the Qb site on the D1 protein, and prevent quinone from binding to this site. Therefore, this group of compounds causes electrons to accumulate on chlorophyll molecules. As a consequence, oxidation reactions in excess of those normally tolerated by the cell occur, and the plant dies. The triazine herbicides (including atrazine) and urea derivatives (diuron) are photosystem II inhibitors.

- Photosystem I inhibitors steal electrons from the normal pathway through FeS to Fdx to NADP leading to direct discharge of electrons on oxygen. As a result, reactive oxygen species are produced and oxidation reactions in excess of those normally tolerated by the cell occur, leading to plant death. Bipyridinium herbicides (such as diquat and paraquat) inhibit the Fe-S – Fdx step of that chain, while diphenyl ether herbicides (such as nitrofen, nitrofluorfen, and acifluorfen) inhibit the Fdx – NADP step.

- HPPD inhibitors inhibit 4-Hydroxyphenylpyruvate dioxygenase, which are involved in tyrosine breakdown. Tyrosine breakdown products are used by plants to make carotenoids, which protect chlorophyll in plants from being destroyed by sunlight. If this happens, the plants turn white due to complete loss of chlorophyll, and the plants die. Mesotrione and sulcotrione are herbicides in this class; a drug, nitisinone, was discovered in the course of developing this class of herbicides.

Herbicide Group (Labeling)

One of the most important methods for preventing, delaying, or managing resistance is to reduce the reliance on a single herbicide mode of action. To do this, farmers must know the mode of action for the herbicides they intend to use, but the relatively complex nature of plant biochemistry makes this difficult to determine. Attempts were made to simplify the understanding of herbicide mode of action by developing a classification system that grouped herbicides by mode of action. Eventually the Herbicide Resistance Action Committee (HRAC) and the Weed Science Society of America (WSSA) developed a classification system. The WSSA and HRAC systems differ in the group designation. Groups in the WSSA and the HRAC systems are designated by numbers and letters, respectively. The goal for adding the "Group" classification and mode of action to the herbicide product label is to provide a simple and practical approach to deliver the information to users. This information will make it easier to develop educational material that is consistent and effective. It should increase user's awareness of herbicide mode of action and provide more accurate recommendations for resistance management. Another goal is to make it easier for users to keep records on which herbicide mode of actions are being used on a particular field from year to year.

Chemical Family

Detailed investigations on chemical structure of the active ingredients of the registered herbicides showed that some moieties (moiety is a part of a molecule that may include either whole functional groups or parts of functional groups as substructures; a functional group has similar chemical properties whenever it occurs in different compounds) have the same mechanisms of action. According to Forouzesh *et al.* 2015, these moieties have been assigned to the names of chemical families and active ingredients are then classified within the chemical families accordingly. Knowing about herbicide chemical family grouping could serve as a short-term strategy for managing resistance to site of action.

Use and Application

Herbicides being sprayed from the spray arms of a tractor in North Dakota.

Most herbicides are applied as water-based sprays using ground equipment. Ground equipment varies in design, but large areas can be sprayed using self-propelled sprayers equipped with long booms, of 60 to 120 feet (18 to 37 m) with spray nozzles spaced every 20–30 inches (510–760 mm) apart. Towed, handheld, and even horse-drawn sprayers are also used. On large areas, herbicides may also at times be applied aerially using helicopters or airplanes, or through irrigation systems (known as chemigation).

A further method of herbicide application developed around 2010, involves ridding the soil of its active weed seed bank rather than just killing the weed. This can successfully treat annual plants but not perennials. Researchers at the Agricultural Research Service found that the application of herbicides to fields late in the weeds' growing season greatly reduces their seed production, and therefore fewer weeds will return the following season. Because most weeds are annuals, their seeds will only survive in soil for a year or two, so this method will be able to destroy such weeds after a few years of herbicide application.

Weed-wiping may also be used, where a wick wetted with herbicide is suspended from a boom and dragged or rolled across the tops of the taller weed plants. This allows treatment of taller grassland weeds by direct contact without affecting related but desirable shorter plants in the grassland sward beneath. The method has the benefit of avoiding spray drift. In Wales, a scheme offering free weed-wiper hire was launched in 2015 in an effort to reduce the levels of MCPA in water courses.

Misuse and Misapplication

Herbicide volatilisation or spray drift may result in herbicide affecting neighboring fields or plants, particularly in windy conditions. Sometimes, the wrong field or plants may be sprayed due to error.

Use Politically, Militarily, and in Conflict

Health and Environmental Effects

Herbicides have widely variable toxicity in addition to acute toxicity from occupational exposure levels.

Some herbicides cause a range of health effects ranging from skin rashes to death. The pathway of attack can arise from intentional or unintentional direct consumption, improper application resulting in the herbicide coming into direct contact with people or wildlife, inhalation of aerial sprays, or food consumption prior to the labeled preharvest interval. Under some conditions, certain herbicides can be transported via leaching or surface runoff to contaminate groundwater or distant surface water sources. Generally, the conditions that promote herbicide transport include intense storm events (particularly shortly after application) and soils with limited capacity to adsorb or retain the herbicides. Herbicide properties that increase likelihood of transport include persistence (resistance to degradation) and high water solubility.

Phenoxy herbicides are often contaminated with dioxins such as TCDD; research has suggested such contamination results in a small rise in cancer risk after occupational exposure to these herbicides. Triazine exposure has been implicated in a likely relationship to increased risk of breast cancer, although a causal relationship remains unclear.

Herbicide manufacturers have at times made false or misleading claims about the safety of their products. Chemical manufacturer Monsanto Company agreed to change its advertising after pressure from New York attorney general Dennis Vacco; Vacco complained about misleading claims that its spray-on glyphosate-based herbicides, including Roundup, were safer than table salt and "practically non-toxic" to mammals, birds, and fish (though proof that this was ever said is hard to find). Roundup is toxic and has resulted in death after being ingested in quantities ranging from 85 to 200 ml, although it has also been ingested in quantities as large as 500 ml with only mild or moderate symptoms. The manufacturer of Tordon 101 (Dow AgroSciences, owned by the Dow Chemical Company) has claimed Tordon 101 has no effects on animals and insects, in spite of evidence of strong carcinogenic activity of the active ingredient Picloram in studies on rats.

The risk of Parkinson's disease has been shown to increase with occupational exposure to herbicides and pesticides. The herbicide paraquat is suspected to be one such factor.

All commercially sold, organic and nonorganic herbicides must be extensively tested prior to approval for sale and labeling by the Environmental Protection Agency. However, because of the large number of herbicides in use, concern regarding health effects is significant. In addition to health effects caused by herbicides themselves, commercial herbicide mixtures often contain other chemicals, including inactive ingredients, which have negative impacts on human health.

Ecological Effects

Commercial herbicide use generally has negative impacts on bird populations, although the impacts are highly variable and often require field studies to predict accurately. Laboratory studies have at times overestimated negative impacts on birds due to toxicity, predicting serious problems that were not observed in the field. Most observed effects are due not to toxicity, but to habitat changes and the decreases in abundance of species on which birds rely for food or shelter. Herbicidc usc in silviculture, used to favor certain types of growth following clearcutting, can cause significant drops in bird populations. Even when herbicides which have low toxicity to birds are used, they decrease the abundance of many types of vegetation on which the birds rely. Herbicide use in agriculture in Britain has been linked to a decline in seed-eating bird species which rely on the weeds killed by the herbicides. Heavy use of herbicides in neotropical agricultural areas has been one of many factors implicated in limiting the usefulness of such agricultural land for wintering migratory birds.

Frog populations may be affected negatively by the use of herbicides as well. While some studies have shown that atrazine may be a teratogen, causing demasculinization in male frogs, the U.S. Environmental Protection Agency (EPA) and its independent Scientific Advisory Panel (SAP) examined all available studies on this topic and concluded that "atrazine does not adversely affect amphibian gonadal development based on a review of laboratory and field studies."

Scientific Uncertainty of Full Extent of Herbicide Effects

The health and environmental effects of many herbicides is unknown, and even the scientific community often disagrees on the risk. For example, a 1995 panel of 13 scientists reviewing studies on the carcinogenicity of 2,4-D had divided opinions on the likelihood 2,4-D causes cancer in humans. As of 1992, studies on phenoxy herbicides were too few to accurately assess the risk of many types of cancer from these herbicides, even though evidence was stronger that exposure to these herbicides is associated with increased risk of soft tissue sarcoma and non-Hodgkin lymphoma. Furthermore, there is some suggestion that herbicides can play a role in sex reversal of certain organisms that experience temperature-dependent sex determination, which could theoretically alter sex ratios.

Resistance

Weed resistance to herbicides has become a major concern in crop production worldwide. Resistance to herbicides is often attributed to lack of rotational programmes of herbicides and to continuous applications of herbicides with the same sites of action. Thus, a true understanding of the sites of action of herbicides is essential for strategic planning of herbicide-based weed control.

Plants have developed resistance to atrazine and to ALS-inhibitors, and more recently, to glyphosate herbicides. Marestail is one weed that has developed glyphosate resistance. Glyphosate-resistant weeds are present in the vast majority of soybean, cotton and corn farms in some U.S. states. Weeds that can resist multiple other herbicides are spreading. Few new herbicides are near commercialization, and none with a molecular mode of action for which there is no resistance. Because most herbicides could not kill all weeds, farmers rotated crops and herbicides to stop resistant weeds. During its initial years, glyphosate was not subject to resistance and allowed farmers to reduce the use of rotation.

A family of weeds that includes waterhemp (Amaranthus rudis) is the largest concern. A 2008-9 survey of 144 populations of waterhemp in 41 Missouri counties revealed glyphosate resistance in 69%. Weeds from some 500 sites throughout Iowa in 2011 and 2012 revealed glyphosate resistance in approximately 64% of waterhemp samples. The use of other killers to target "residual" weeds has become common, and may be sufficient to have stopped the spread of resistance From 2005 through

2010 researchers discovered 13 different weed species that had developed resistance to glyphosate. But since then only two more have been discovered. Weeds resistant to multiple herbicides with completely different biological action modes are on the rise. In Missouri, 43% of samples were resistant to two different herbicides; 6% resisted three; and 0.5% resisted four. In Iowa 89% of waterhemp samples resist two or more herbicides, 25% resist three, and 10% resist five.

For southern cotton, herbicide costs has climbed from between $50 and $75 per hectare a few years ago to about $370 per hectare in 2013. Resistance is contributing to a massive shift away from growing cotton; over the past few years, the area planted with cotton has declined by 70% in Arkansas and by 60% in Tennessee. For soybeans in Illinois, costs have risen from about $25 to $160 per hectare.

Dow, Bayer CropScience, Syngenta and Monsanto are all developing seed varieties resistant to herbicides other than glyphosate, which will make it easier for farmers to use alternative weed killers. Even though weeds have already evolved some resistance to those herbicides, Powles says the new seed-and-herbicide combos should work well if used with proper rotation.

Biochemistry of Resistance

Resistance to herbicides can be based on one of the following biochemical mechanisms:

- Target-site resistance: This is due to a reduced (or even lost) ability of the herbicide to bind to its target protein. The effect usually relates to an enzyme with a crucial function in a metabolic pathway, or to a component of an electron-transport system. Target-site resistance may also be caused by an overexpression of the target enzyme (via gene amplification or changes in a gene promoter).

- Non-target-site resistance: This is caused by mechanisms that reduce the amount of herbicidal active compound reaching the target site. One important mechanism is an enhanced metabolic detoxification of the herbicide in the weed, which leads to insufficient amounts of the active substance reaching the target site. A reduced uptake and translocation, or sequestration of the herbicide, may also result in an insufficient herbicide transport to the target site.

- Cross-resistance: In this case, a single resistance mechanism causes resistance to several herbicides. The term target-site cross-resistance is used when the herbicides bind to the same target site, whereas non-target-site cross-resistance is due to a single non-target-site mechanism (e.g., enhanced metabolic detoxification) that entails resistance across herbicides with different sites of action.

- Multiple resistance: In this situation, two or more resistance mechanisms are present within individual plants, or within a plant population.

Resistance Management

Worldwide experience has been that farmers tend to do little to prevent herbicide resistance developing, and only take action when it is a problem on their own farm or neighbor's. Careful observation is important so that any reduction in herbicide efficacy can be detected. This may indicate evolving resistance. It is vital that resistance is detected at an early stage as if it becomes an acute, whole-farm problem, options are more limited and greater expense is almost inevitable. Table 1 lists factors which enable the risk of resistance to be assessed. An essential pre-requisite for confirmation of resistance is a good diagnostic test. Ideally this should be rapid, accurate, cheap and accessible. Many diagnostic tests have been developed, including glasshouse pot assays, petri dish assays and chlorophyll fluorescence. A key component of such tests is that the response of the suspect population to a herbicide can be compared with that of known susceptible and resistant standards under controlled conditions. Most cases of herbicide resistance are a consequence of the repeated use of herbicides, often in association with crop monoculture and reduced cultivation practices. It is necessary, therefore, to modify these practices in order to prevent or delay the onset of resistance or to control existing resistant populations. A key objective should be the reduction in selection pressure. An integrated weed management (IWM) approach is required, in which as many tactics as possible are used to combat weeds. In this way, less reliance is placed on herbicides and so selection pressure should be reduced.

Optimising herbicide input to the economic threshold level should avoid the unnecessary use of herbicides and reduce selection pressure. Herbicides should be used to their greatest potential by ensuring that the timing, dose, application method, soil and climatic conditions are optimal for good activity. In the UK, partially resistant grass weeds such as *Alopecurus myosuroides* (blackgrass) and *Avena* spp. (wild oat) can often be controlled adequately when herbicides are applied at the 2-3 leaf stage, whereas later applications at the 2-3 tiller stage can fail badly. Patch spraying, or applying herbicide to only the badly infested areas of fields, is another means of reducing total herbicide use.

Table 1. Agronomic factors influencing the risk of herbicide resistance development

Factor	Low risk	High risk
Cropping system	Good rotation	Crop monoculture
Cultivation system	Annual ploughing	Continuous minimum tillage
Weed control	Cultural only	Herbicide only
Herbicide use	Many modes of action	Single modes of action
Control in previous years	Excellent	Poor
Weed infestation	Low	High
Resistance in vicinity	Unknown	Common

Approaches to Treating Resistant Weeds

Alternative Herbicides

When resistance is first suspected or confirmed, the efficacy of alternatives is likely to be the first consideration. The use of alternative herbicides which remain effective on resistant populations can be a successful strategy, at least in the short term. The effectiveness of alternative herbicides will be highly dependent on the extent of cross-resistance. If there is resistance to a single group of herbicides, then the use of herbicides from other groups may provide a simple and effective solution, at least in the short term. For example, many triazine-resistant weeds have been readily controlled by the use of alternative herbicides such as dicamba or glyphosate. If resistance extends to more than one herbicide group, then choices are more limited. It should not be assumed that resistance will automatically extend to all herbicides with the same mode of action, although it is wise to assume this until proved otherwise. In many weeds the degree of cross-resistance between the five groups of ALS inhibitors varies considerably. Much will depend on the resistance mechanisms present, and it should not be assumed that these will necessarily be the same in different populations of the same species. These differences are due, at least in part, to the existence of different mutations conferring target site resistance. Consequently, selection for different mutations may result in different patterns of cross-resistance. Enhanced metabolism can affect even closely related herbicides to differing degrees. For example, populations of *Alopecurus myosuroides* (blackgrass) with an enhanced metabolism mechanism show resistance to pendimethalin but not to trifluralin, despite both being dinitroanilines. This is due to differences in the vulnerability of these two herbicides to oxidative metabolism. Consequently, care is needed when trying to predict the efficacy of alternative herbicides.

Mixtures and Sequences

The use of two or more herbicides which have differing modes of action can reduce the selection for resistant genotypes. Ideally, each component in a mixture should:

- Be active at different target sites

- Have a high level of efficacy

- Be detoxified by different biochemical pathways

- Have similar persistence in the soil (if it is a residual herbicide)

- Exert negative cross-resistance

- Synergise the activity of the other component

No mixture is likely to have all these attributes, but the first two listed are the most important. There is a risk that mixtures will select for resistance to both components in the longer term. One practical advantage of sequences of two herbicides compared with mixtures is that a better appraisal of the efficacy of each herbicide component is possible, provided that sufficient time elapses between each application. A disadvantage with sequences is that two separate applications have to be made and it is possible that the later application will be less effective on weeds surviving the first application. If these are resistant, then the second herbicide in the sequence may increase selection for resistant individuals by killing the susceptible plants which were damaged but not killed by the first application, but allowing the larger, less affected, resistant plants to survive. This has been cited as one reason why ALS-resistant *Stellaria media* has evolved in Scotland recently (2000), despite the regular use of a sequence incorporating mecoprop, a herbicide with a different mode of action.

Herbicide Rotations

Rotation of herbicides from different chemical groups in successive years should reduce selection for resistance. This is a key element in most resistance prevention programmes. The value of this approach depends on the extent of cross-resistance, and whether multiple resistance occurs owing to the presence of several different resistance mechanisms. A practical problem can be the lack of awareness by farmers of the different groups of herbicides that exist. In Australia a scheme has been introduced in which identifying letters are included on the product label as a means of enabling farmers to distinguish products with different modes of action.

Farming Practices and Resistance: a Case Study

Herbicide resistance became a critical problem in Australian agriculture, after many Australian sheep farmers began to exclusively grow wheat in their pastures in the 1970s. Introduced varieties of ryegrass, while good for grazing sheep, compete intensely with wheat. Ryegrasses produce so many seeds that, if left unchecked, they can completely choke a field. Herbicides provided excellent control, while reducing soil disrupting because of less need to plough. Within little more than a decade, ryegrass and other weeds began to develop resistance. In response Australian farmers changed methods. By 1983, patches of ryegrass had become immune to Hoegrass, a family of herbicides that inhibit an enzyme called acetyl coenzyme A carboxylase.

Ryegrass populations were large, and had substantial genetic diversity, because farmers had planted many varieties. Ryegrass is cross-pollinated by wind, so genes shuffle frequently. To control its distribution farmers sprayed inexpensive Hoegrass, creating selection pressure. In addition, farmers sometimes diluted the herbicide in order to save money, which allowed some plants to survive application. When resistance appeared farmers turned to a group of herbicides that block acetolactate synthase. Once again, ryegrass in Australia evolved a kind of "cross-resistance" that allowed it to rapidly break down a variety of herbicides. Four classes of herbicides become ineffective within a few years. In 2013 only two herbicide classes, called Photosystem II and long-chain fatty acid inhibitors, were effective against ryegrass.

List of Common Herbicides

Synthetic Herbicides

- 2,4-D is a broadleaf herbicide in the phenoxy group used in turf and no-till field crop production. Now, it is mainly used in a blend with other herbicides to allow lower rates of herbicides to be used; it is the most widely used herbicide in the world, and third most commonly used in the United States. It is an example of synthetic auxin (plant hormone).

- Aminopyralid is a broadleaf herbicide in the pyridine group, used to control weeds on grassland, such as docks, thistles and nettles. It is notorious for its ability to persist in compost.

- Atrazine, a triazine herbicide, is used in corn and sorghum for control of broadleaf weeds and grasses. Still used because of its low cost and because it works well on a broad spectrum of weeds common in the US corn belt, atrazine is commonly used with other herbicides to reduce the overall rate of atrazine and to lower the potential for groundwater contamination; it is a photosystem II inhibitor.

- Clopyralid is a broadleaf herbicide in the pyridine group, used mainly in turf, rangeland, and for control of noxious thistles. Notorious for its ability to persist in compost, it is another example of synthetic auxin.

- Dicamba, a postemergent broadleaf herbicide with some soil activity, is used on turf and field corn. It is another example of a synthetic auxin.

- Glufosinate ammonium, a broad-spectrum contact herbicide, is used to control weeds after the crop emerges or for total vegetation control on land not used for cultivation.

- Fluazifop (Fuselade Forte), a post emergence, foliar absorbed, translocated grass-selective herbicide with little residual action. It is used on a very wide range of broad leaved crops for control of annual and perennial grasses.

- Fluroxypyr, a systemic, selective herbicide, is used for the control of broad-leaved weeds in small grain cereals, maize, pastures, rangeland and turf. It is a synthetic auxin. In cereal growing, fluroxypyr's key importance is control of cleavers, *Galium aparine*. Other key broadleaf weeds are also controlled.

- Glyphosate, a systemic nonselective herbicide, is used in no-till burndown and for weed control in crops genetically modified to resist its effects. It is an example of an EPSPs inhibitor.

- Imazapyr a nonselective herbicide, is used for the control of a broad range of weeds, including terrestrial annual and perennial grasses and broadleaf herbs, woody species, and riparian and emergent aquatic species.

- Imazapic, a selective herbicide for both the pre- and postemergent control of some annual and perennial grasses and some broadleaf weeds, kills plants by inhibiting the production of branched chain amino acids (valine, leucine, and isoleucine), which are necessary for protein synthesis and cell growth.

- Imazamox, an imidazolinone manufactured by BASF for postemergence application that is an acetolactate synthase (ALS) inhibitor. Sold under trade names Raptor, Beyond, and Clearcast.

- Linuron is a nonselective herbicide used in the control of grasses and broadleaf weeds. It works by inhibiting photosynthesis.

- MCPA (2-methyl-4-chlorophenoxyacetic acid) is a phenoxy herbicide selective for broadleaf plants and widely used in cereals and pasture.

- Metolachlor is a pre-emergent herbicide widely used for control of annual grasses in corn and sorghum; it has displaced some of the atrazine in these uses.

- Paraquat is a nonselective contact herbicide used for no-till burndown and in aerial destruction of marijuana and coca plantings. It is more acutely toxic to people than any other herbicide in widespread commercial use.

- Pendimethalin, a pre-emergent herbicide, is widely used to control annual grasses and some broad-leaf weeds in a wide range of crops, including corn, soybeans, wheat, cotton, many tree and vine crops, and many turfgrass species.

- Picloram, a pyridine herbicide, mainly is used to control unwanted trees in pastures and edges of fields. It is another synthetic auxin.

- Sodium chlorate *(disused/banned in some countries)*, a nonselective herbicide, is considered phytotoxic to all green plant parts. It can also kill through root absorption.

- Triclopyr, a systemic, foliar herbicide in the pyridine group, is used to control broadleaf weeds while leaving grasses and conifers unaffected.

- Several sulfonylureas, including Flazasulfuron and Metsulfuron-methyl, which act as ALS inhibitors and in some cases are taken up from the soil via the roots.

Organic Herbicides

Recently, the term "organic" has come to imply products used in organic farming. Under this definition, an organic herbicide is one that can be used in a farming enterprise that has been classified as organic. Commercially sold organic herbicides are expensive and may not be affordable for commercial farming. Depending on the application, they may be less effective than synthetic herbicides and are generally used along with cultural and mechanical weed control practices.

Homemade organic herbicides include:

- Corn gluten meal (CGM) is a natural pre-emergence weed control used in turfgrass, which reduces germination of many broadleaf and grass weeds.

- Vinegar is effective for 5–20% solutions of acetic acid, with higher concentrations most effective, but it mainly destroys surface growth, so respraying to treat regrowth is needed. Resistant plants generally succumb when weakened by respraying.

- Steam has been applied commercially, but is now considered uneconomical and inadequate. It controls surface growth but not underground growth and so respraying to treat regrowth of perennials is needed.

- Flame is considered more effective than steam, but suffers from the same difficulties.

- D-limonene (citrus oil) is a natural degreasing agent that strips the waxy skin or cuticle from weeds, causing dehydration and ultimately death.

- Saltwater or salt applied in appropriate strengths to the rootzone will kill most plants.

- Monocerin produced by certain fungi will kill certain weeds such as Johnson grass.

Of Historical Interest and Other

- 2,4,5-Trichlorophenoxyacetic acid (2,4,5-T) was a widely used broadleaf herbicide until being phased out starting in the late 1970s. While 2,4,5-T itself is of only moderate toxicity, the manufacturing process for 2,4,5-T contaminates this chemical with trace amounts of 2,3,7,8-tetrachlorodibenzo-p-dioxin (TCDD). TCDD is extremely toxic to humans. With proper temperature control during production of 2,4,5-T, TCDD levels can be held to about .005 ppm. Before the TCDD risk was well understood, early production facilities lacked proper temperature controls. Individual batches tested later were found to have as much as 60 ppm of TCDD. 2,4,5-T was withdrawn from use in the USA in 1983, at a time of heightened public sensitivity about chemical hazards in the environment. Public concern about dioxins was high, and production and use of other (non-herbicide) chemicals potentially containing TCDD contamination was also withdrawn. These included pentachlorophenol (a wood preservative) and PCBs (mainly used as stabilizing agents in transformer oil). Some feel that the 2,4,5-T withdrawal was not based on sound science. 2,4,5-T has since largely been replaced by dicamba and triclopyr.

- Agent Orange was a herbicide blend used by the British military during the Malayan Emergency and the U.S. military during the Vietnam War between January 1965 and April 1970 as a defoliant. It was a 50/50 mixture of the *n*-butyl esters of 2,4,5-T and 2,4-D. Because of TCDD contamination in the 2,4,5-T component, it has been blamed for serious illnesses in many people who were exposed to it. However, research on populations exposed to its dioxin contaminant have been inconsistent and inconclusive.

- Diesel, and other heavy oil derivatives, are known to be informally used at times, but are usually banned for this purpose.

Liming (Soil)

Liming is the application of calcium- and magnesium-rich materials to soil in various forms, including marl, chalk, limestone, or hydrated lime. This neutralises soil acidity and increases activity of soil bacteria. However, oversupply may result in harm to plant life. Lime is a basic chemical, the effect of it makes the soil more basic thus making acidic soils neutral.

Liming of a field in Devon

The degree to which a given amount of lime per unit of soil volume will increase soil pH depends on the cation exchange capacity (CEC). Soils with low CEC will show a more marked pH increase than soils with high CEC. But the low-CEC soils will witness more rapid leaching of the added bases, and so will see a quicker return to original acidity unless additional liming is done.

Over-liming is most likely to occur on soil which has low CEC, such as sand which is deficient in buffering agents such as organic matter and clay.

Most acid soils are saturated with aluminium rather than hydrogen ions. The acidity of the soil is therefore a result of hydrolysis of aluminium. This concept of "corrected lime potential" to define the degree of base saturation in soils became the basis for procedures now used in soil testing laboratories to determine the "lime requirement" of soils.

An agricultural study at the Faculty of Forestry in Freising, Germany that compared tree stocks 2 and 20 years after liming found that liming promotes nitrate leaching and decreases the phosphorus content of some leaves.

Animal Feed

A photo of a feedlot in Texas, USA, where cattle are "finished" (fattened on grains) prior to slaughter.

Animal feed is food given to domestic animals in the course of animal husbandry. There are two basic types, *fodder* and *forage*. Used alone, the word "feed" more often refers to fodder.

Fodder

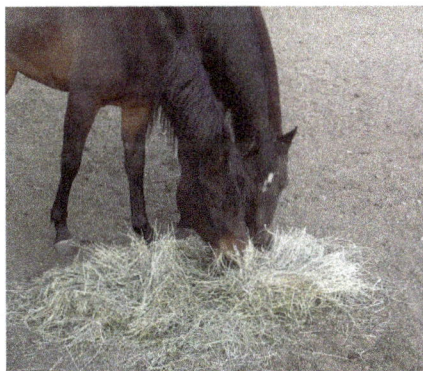

Equine nutritionists recommend that 50% or more of a horse's diet by weight should be forages, such as hay

"Fodder" refers particularly to foods or forages given to the animals (including plants cut and carried to them), rather than that which they forage for themselves. It includes hay, straw, silage, compressed and pelleted feeds, oils and mixed rations, and sprouted grains and legumes. Feed grains are the most important source of animal feed globally. The amount of grain used to produce the same unit of meat varies substantially. According to an estimate reported by the BBC in 2008, "Cows and sheep need 8kg of grain for every 1kg of meat they produce, pigs about 4kg. The most efficient poultry units need a mere 1.6kg of feed to produce 1kg of chicken." Farmed fish can also be fed on grain, and use even less than poultry. The two most important feed grains are maize and soyabean, and the United States is by far the largest exporter of both, averaging about half of the global maize trade and 40% of the global soya trade in the years leading up the 2012 drought. Other feed grains include wheat, oats, barley, and rice, among many others.

Traditional sources of animal feed include household food scraps and the byproducts of food processing industries such as milling and brewing. Material remaining from milling oil crops like peanuts, soy, and corn are important sources of fodder. Scraps fed to pigs are called slop, and those fed to chicken are called chicken scratch. Brewer's spent grain is a byproduct of beer making that is widely used as animal feed.

A pelleted ration designed for horses

Compound feed is fodder that is blended from various raw materials and additives. These blends are formulated according to the specific requirements of the target animal. They are manufactured by feed compounders as *meal type*, *pellets* or *crumbles*. The main ingredients used in commercially prepared feed are the feed grains, which include corn, soybeans, sorghum, oats, and barley.

Compound feed may also include premixes, which may also be sold separately. Premixes are composed of microingredients such as vitamins, minerals, chemical preservatives, antibiotics, fermentation products, and other essential ingredients that are purchased from premix companies, usually in sacked form, for blending into commercial rations. Because of the availability of these products, a farmer who uses his own grain can formulate his own rations and be assured his animals are getting the recommended levels of minerals and vitamins.

According to the American Feed Industry Association, as much as $20 billion worth of feed ingredients are purchased each year. These products range from grain mixes to orange rinds to beet pulps. The feed industry is one of the most competitive businesses in the agricultural sector, and is by far the largest purchaser of U.S. corn, feed grains, and soybean meal. Tens of thousands of farmers with feed mills on their own farms are able to compete with huge conglomerates with national distribution. Feed crops generated $23.2 billion in cash receipts on U.S. farms in 2001. At the same time, farmers spent a total of $24.5 billion on feed that year.

In 2011, around 734.5 million tons of feed were produced annually around the world.

History

The beginning of industrial-scale production of animal feeds can be traced back to the late 19th century, around the time advances in human and animal nutrition were able to identify the benefits of a balanced diet, and the importance of the role processing of certain raw materials played. Corn gluten feed was first manufactured in 1882, while leading world feed producer Purina Feeds was established in 1894 by William Hollington Danforth. Cargill, which was mainly dealing in grains from its beginnings in 1865, started to deal in feed at about 1884.

The feed industry expanded rapidly in the first quarter of the 20th century, with Purina expanding its operations into Canada, and opened its first feed mill in 1927 (which is still in operation). In 1928, the feed industry was revolutionized by the introduction of the first pelleted feeds - Purina Checkers.

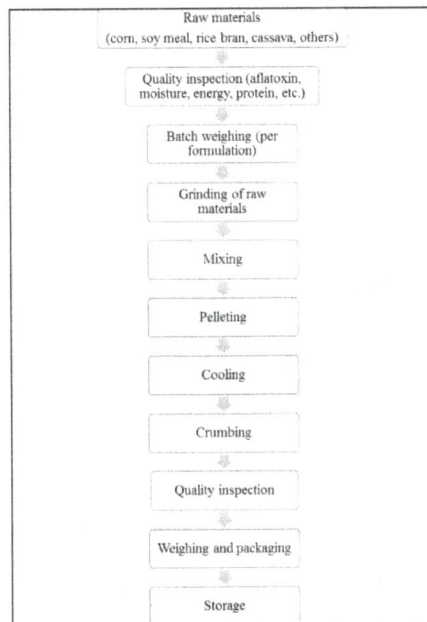

Workflow for general feed manufacturing process

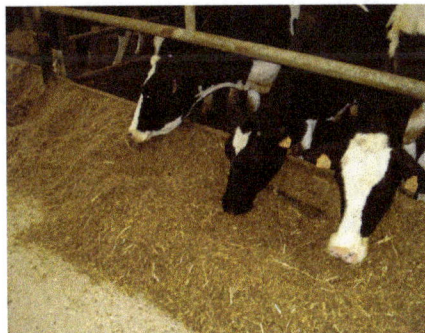

Cattle eating a total mixed ration

Manufacturers

The job of the feed manufacturer is to buy the commodities and blend them in the feed mill according to the specifications outlined by the animal nutritionist. There is little room for error because, if the ration is not apportioned correctly, lowered animal production and diminished outward appearance can occur.

Asia

One of the largest Asian feed producers is Charoen Pokphand (the CP Group), a Thai company producing 18 million tonnes of compound feed at various locations across East Asia.

Europe

The merge of the Hamburg-based traditional commodity trade firm, Cremer, and the Düsseldorf based Deuka (Deutsche Kraftfutterwerke), led to one of the largest feed companies in Europe. The new Cremer Group produces around 3.5 million tons. BOCM Pauls in the UK produces around the same amount if not more.

United States

Feed production facility in Oneonta, New York

Leading U.S. companies involved in prepared feeds production in the early first decade of the 21st century included ConAgra Inc., an Omaha, Nebraska-based firm; and Cargill, Incorporated, a diversified company that was the nation's top exporter of grain. In 1998, Ralston Purina Company, based in St. Louis, Missouri, formed Agribrands International, Inc. to control its international animal feed and agricultural products division. Agribrands produced feed and other products for livestock in markets outside of the United States, and had about 75 facilities operating in 16 countries. In 2001, it was acquired by Cargill.

Other significant industry players included Conti Group Companies, Inc., the world's leading cattle feeder; CHS, Inc. (previously known as Cenex Harvest States Cooperative), which was primarily involved in grain trading; and Farmland Industries, Inc., the leading agricultural cooperative in the United States. Farmland was a worldwide exporter of products, such as grain. In May 2002, the firm declared bankruptcy, and in the following year, Smithfield Foods acquired most of Farmland's assets.

Forage

"Forage" is plant material (mainly plant leaves and stems) eaten by grazing livestock. Historically, the term *forage* has meant only plants eaten by the animals directly as pasture, crop residue, or immature cereal crops, but it is also used more loosely to include similar plants cut for fodder and carried to the animals, especially as hay or silage.

In this photo a herdsman from the Masaai people watches as his cattle graze in the Ngorongoro crater, Tanzania.

Manufacture

Nutrition

In agriculture today, the nutritional needs of farm animals are well understood and may be satisfied through natural forage and fodder alone, or augmented by direct supplementation of nutrients in concentrated, controlled form. The nutritional quality of feed is influenced not only by the nutrient content, but also by many other factors such as feed presentation, hygiene, digestibility, and effect on intestinal health.

Feed additives provide a mechanism through which these nutrient deficiencies can be resolved effect the rate of growth of such animals and also their health and well-being. Even with all the benefits of higher quality feed, most of a farm animal's diet still consists of grain-based ingredients because of the higher costs of quality feed.

By Animal

- Bird food
- Cat food
- Cattle feeding
- Dog food
- Equine nutrition
- Pet food
- Pig farming
- Poultry feed
- Sheep husbandry

Mechanised Agriculture

Mechanised agriculture is the process of using agricultural machinery to mechanise the work of agriculture, greatly increasing farm worker productivity. In modern times, powered machinery has replaced many farm jobs formerly carried out by manual labour or by working animals such as oxen, horses and mules.

A cotton picker at work. The first successful models were introduced in the mid-1940s and each could do the work of 50 hand pickers.

The entire history of agriculture contains many examples of the use of tools, such as the hoe and the plough. But the ongoing integration of machines since the Industrial Revolution has allowed farming to become much less labour-intensive.

Current mechanised agriculture includes the use of tractors, trucks, combine harvesters, countless types of farm implements, aeroplanes and helicopters (for aerial application), and other vehicles. Precision agriculture even uses computers in conjunction with satellite imagery and satellite navigation (GPS guidance) to increase yields.

Mechanisation was one of the large factors responsible for urbanisation and industrial economies. Besides improving production efficiency, mechanisation encourages large scale production and sometimes can improve the quality of farm produce. On the other hand, it can displace unskilled farm labour and can cause environmental degradation (such as pollution, deforestation, and soil erosion), especially if it is applied shortsightedly rather than holistically.

History

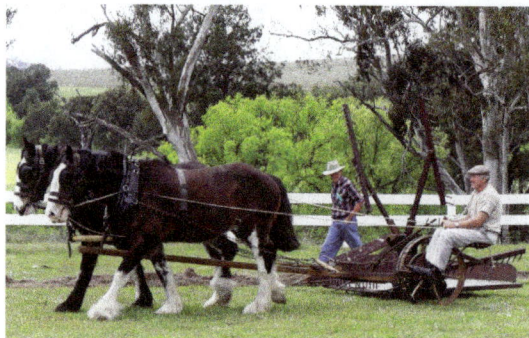

A reaper at Woolbrook, New South Wales

Jethro Tull's seed drill (ca. 1701) was a mechanical seed spacing and depth placing device that increased crop yields and saved seed. It was an important factor in the British Agricultural Revolution.

Threshing machine in 1881. Steam engines were also used to power threshing machines. Today both reaping and threshing are done with a combine harvester.

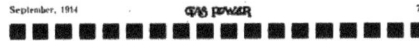

"Better and cheaper than horses" was the theme of many advertisements of the 1910s through 1930s.

Since the beginning of agriculture threshing was done by hand with a flail, requiring a great deal of labour. The threshing machine, which was invented in 1794 but not widely used for several more decades, simplified the operation and allowed the use of animal power. Before the invention of the grain cradle (ca. 1790) an able bodied labourer could reap about one quarter acre of wheat in a day using a sickle. It was estimated that for each of Cyrus McCormick's horse pulled reapers (ca. 1830s) freed up five men for military service in the US Civil War. Later innovations included raking and binding machines. By 1890 two men and two horses could cut, rake and bind 20 acres of wheat per day.

In the 1880s the reaper and threshing machine were combined into the combine harvester. These machines required large teams of horses or mules to pull. Steam power was applied to threshing machines in the late 19th century. There were steam engines that moved around on wheels under their own power for supplying temporary power to stationary threshing machines. These were called *road engines,* and Henry Ford seeing one as a boy was inspired to build an automobile.

With internal combustion came the first modern tractors in the early 1900s, becoming more popular after the Fordson tractor (ca. 1917). At first reapers and combine harvesters were pulled by tractors, but in the 1930s self powered combines were developed. (*Link to a chapter on agricultural mechanisation in the 20th Century at reference*)

Advertising for motorised equipment in farm journals during this era did its best to compete against horse-drawn methods with economic arguments, extolling common themes such as that a tractor "eats only when it works", that one tractor could replace many horses, and that mechanisation could allow one man to get more work done per day than he ever had before. The horse population in the US began to decline in the 1920s after the conversion of agriculture and transportation to internal combustion. Peak tractor sales in the US were around 1950. In addition to saving labour, this freed up much land previously used for supporting draft animals. The greatest period of growth in agricultural productivity in the US was from the 1940s to the 1970s, during which time agriculture was benefiting from internal combustion powered tractors and combine harvesters, chemical fertilizers and the green revolution.

Although farmers of corn, wheat, soy, and other commodity crops had replaced most of their workers with harvesting machines and combines enabling them to efficiently cut and gather grains, growers of produce continued to rely on human pickers to avoid the bruising of the product in order to maintain the blemish-free appearance demanded of consumers. The continuous supply of illegal workers from Latin America that were willing to harvest the crops for low wages further suppressed the need for mechanisation. As the number of illegal workers has continued to decline since reaching its peak in 2007 due to increased border patrols and an improving Mexican economy, the industry is increasing the use of mechanisation. Proponents argue that mechanisation will boost productivity and help to maintain low food prices while farm worker advocates assert that it will eliminate jobs and will give an advantage to large growers who are able to afford the required equipment.

Current Status of Future Applications

Asparagus Harvesting

Asparagus are presently harvested by hand with labour costs at 71% of production costs and 44% of selling costs. Asparagus is a difficult crop to harvest since each spear matures at a different speed making it difficult to achieve a uniform harvest. A prototype asparagus harvesting machine - using a light-beam sensor to identify the taller spears - is expected to be available for commercial use.

Blueberry Harvesting

Mechanization of Maine's blueberry industry has reduced the number of migrant workers required from 5,000 in 2005 to 1,500 in 2015 even though production has increased from 50-60 million pounds per year in 2005 to 90 million pounds in 2015.

Chili Pepper Harvesting

As of 2014, prototype chili pepper harvesters are being tested by New Mexico State University. The New Mexico green chile crop is currently hand-picked entirely by field workers as chili pods tend to bruise easily. The first commercial application commenced in 2015. The equipment is expected to increase yield per acre and help to offset a sharp decline in acreage planted due to the lack of available labour and drought conditions.

Orange Harvesting

As of 2010, approximately 10% of the processing orange acreage in Florida is harvested mechanically. Mechanisation has progressed slowly due to the uncertainty of future economic benefits due to competition from Brazil and the transitory damage to orange trees when they are harvested.

Raisin Harvesting

As of 2007, mechanised harvesting of raisins is at 45%; however the rate has slowed due to high raisin demand and prices making the conversion away from hand labour less urgent. A new strain of grape developed by the USDA that drys on the vine and is easily harvested mechanically is expected to reduce the demand for labour.

Strawberry Harvesting

Strawberries are a high cost-high value crop with the economics supporting mechanisation. In 2005, picking and hauling costs were estimated at $594 per ton or 51% of the total grower cost. However, the delicate nature of fruit make it an unlikely candidate for mechanisation in the near future. A strawberry harvester developed by Shibuya Seiki and unveiled in Japan in 2013 is able to pick a strawberry every eight seconds. The robot identifies which strawberries are ready to pick by using three separate cameras and then once identified as ready, a mechanised arm snips the fruit free and gently places it in a basket. The robot moves on rails between the rows of strawberries which are generally contained within elevated greenhouses. The machine costs 5 million yen. A new strawberry harvester made by Agrobot that will harvest strawberries on raised, hydroponic beds using 60 robotic arms is expected to be released in 2016.

Tomato Harvesting

Mechanical harvesting of tomatoes started in 1965 and as of 2010, nearly all processing tomatoes are mechanically harvested. As of 2010, 95% of the US processed tomato crop is produced in California. Although fresh market tomatoes have substantial hand harvesting costs (in 2007, the costs of hand picking and hauling were $86 per ton which is 19% of total grower cost), packing and selling costs were more of a concern (at 44% of total grower cost) making it likely that cost saving efforts would be applied there.

According to a 1977 report by the California Agrarian Action Project, during the summer of 1976 in California, many harvest machines had been equipped with a photo-electric scanner that sorted out green tomatoes among the ripe red ones using infrared lights and colour sensors. It worked in lieu of 5,000 hand harvesters causing displacement of innumerable farm labourers as well as wage cuts and shorter work periods. Migrant workers were hit the hardest. To withstand the rigour of the machines, new crop varieties were bred to match the automated pickers. UC Davis Professor G.C. Hanna propagated a thick-skinned tomato called VF-145. But even still, millions were damaged with impact cracks and university breeders produced a more tougher and juiceless "square round" tomato. Small farms were of insufficient size to obtain financing to purchase the equipment and within 10 years, 85% of the state's 4,000 cannery tomato farmers were out of the business. This led to a concentrated tomato industry in California that "now packed 85% of the nation's tomato products". The monoculture fields fostered rapid pest growth, requiring the use of "more than four million pounds of pesticides each year" which greatly affected the health of the soil, the farm workers, and possibly the consumers.

Irrigation

An irrigation sprinkler watering a lawn

Irrigation canal in Osmaniye, Turkey

Irrigation is the method in which a controlled amount of water is supplied to plants at regular intervals for agriculture. It is used to assist in the growing of agricultural crops, maintenance of landscapes, and revegetation of disturbed soils in dry areas and during periods of inadequate rainfall. Additionally, irrigation also has a few other uses in crop production, which include protecting plants against frost, suppressing weed growth in grain fields and preventing soil consolidation. In contrast, agriculture that relies only on direct rainfall is referred to as rain-fed or dry land farming.

Irrigation systems are also used for dust suppression, disposal of sewage, and in mining. Irrigation is often studied together with drainage, which is the natural or artificial removal of surface and sub-surface water from a given area.

Irrigation has been a central feature of agriculture for over 5,000 years and is the product of many cultures. Historically, it was the basis for economies and societies across the globe, from Asia to the Southwestern United States.

History

Animal-powered irrigation, Upper Egypt, ca. 1846

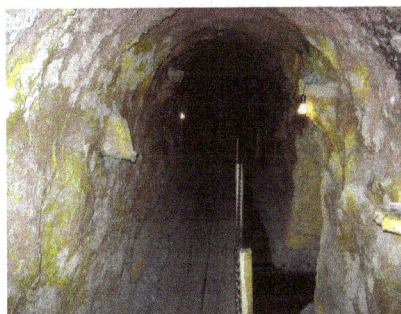

Inside a karez tunnel at Turpan, Sinkiang

ArchaMIhceal ified as evidence of irrigation where the natural rainfall was insufficient to support crops for rainfed agriculture.

Perennial irrigation was practiced in the Mesopotamian plain whereby crops were regularly watered throughout the growing season by coaxing water through a matrix of small channels formed in the field.

irrigation in Tamil Nadu (India)

Ancient Egyptians practiced *Basin irrigation* using the flooding of the Nile to inundate land plots which had been surrounded by dykes. The flood water was held until the fertile sediment had settled before the surplus was returned to the watercourse. There is evidence of the ancient Egyptian pharaoh Amenemhet III in the twelfth dynasty (about 1800 BCE) using the natural lake of the Faiyum Oasis as a reservoir to store surpluses of water for use during the dry seasons, the lake swelled annually from flooding of the Nile.

The Ancient Nubians developed a form of irrigation by using a waterwheel-like device called a *sakia*. Irrigation began in Nubia some time between the third and second millennium BCE. It largely depended upon the flood waters that would flow through the Nile River and other rivers in what is now the Sudan.

In sub-Saharan Africa irrigation reached the Niger River region cultures and civilizations by the first or second millennium BCE and was based on wet season flooding and water harvesting.

Terrace irrigation is evidenced in pre-Columbian America, early Syria, India, and China. In the Zana Valley of the Andes Mountains in Peru, archaeologists found remains of three irrigation canals radiocarbon dated from the 4th millennium BCE, the 3rd millennium BCE and the 9th century CE. These canals are the earliest record of irrigation in the New World. Traces of a canal possibly dating from the 5th millennium BCE were found under the 4th millennium canal. Sophisticated irrigation and storage systems were developed by the Indus Valley Civilization in present-day Pakistan and North India, including the reservoirs at Girnar in 3000 BCE and an early canal irrigation system from circa 2600 BCE. Large scale agriculture was practiced and an extensive network of canals was used for the purpose of irrigation.

Ancient Persia (modern day Iran) as far back as the 6th millennium BCE, where barley was grown in areas where the natural rainfall was insufficient to support such a crop. The Qanats, developed in ancient Persia in about 800 BCE, are among the oldest known irrigation methods still in use today. They are now found in Asia, the Middle East and North Africa. The system comprises a network of vertical wells and gently sloping tunnels driven into the sides of cliffs and steep hills to tap groundwater. The noria, a water wheel with clay pots around the rim powered by the flow of the stream (or by animals where the water source was still), was first brought into use at about this time, by Roman settlers in North Africa. By 150 BCE the pots were fitted with valves to allow smoother filling as they were forced into the water.

The irrigation works of ancient Sri Lanka, the earliest dating from about 300 BCE, in the reign of King Pandukabhaya and under continuous development for the next thousand years, were one of the most complex irrigation systems of the ancient world. In addition to underground canals, the Sinhalese were the first to build completely artificial reservoirs to store water. Due to their engineering superiority in this sector, they were often called 'masters of irrigation'.Most of these irrigation systems still exist undamaged up to now, in Anuradhapura and Polonnaruwa, because of the advanced and precise engineering. The system was extensively restored and further extended during the reign of King Parakrama Bahu (1153–1186 CE).

China

The oldest known hydraulic engineers of China were Sunshu Ao (6th century BCE) of the Spring and Autumn Period and Ximen Bao (5th century BCE) of the Warring States period, both of whom worked on large irrigation projects. In the Sichuan region belonging to the State of Qin of ancient

China, the Dujiangyan Irrigation System was built in 256 BCE to irrigate an enormous area of farmland that today still supplies water. By the 2nd century AD, during the Han Dynasty, the Chinese also used chain pumps that lifted water from lower elevation to higher elevation. These were powered by manual foot pedal, hydraulic waterwheels, or rotating mechanical wheels pulled by oxen. The water was used for public works of providing water for urban residential quarters and palace gardens, but mostly for irrigation of farmland canals and channels in the fields.

Korea

In 15th century Korea, the world's first rain gauge, *uryanggye*, was invented in 1441. The inventor was Jang Yeong-sil, a Korean engineer of the Joseon Dynasty, under the active direction of the king, Sejong the Great. It was installed in irrigation tanks as part of a nationwide system to measure and collect rainfall for agricultural applications. With this instrument, planners and farmers could make better use of the information gathered in the survey.

North America

In North America, the Hohokam were the only culture known to rely on irrigation canals to water their crops, and their irrigation systems supported the largest population in the Southwest by AD 1300. The Hohokam constructed an assortment of simple canals combined with weirs in their various agricultural pursuits. Between the 7th and 14th centuries, they also built and maintained extensive irrigation networks along the lower Salt and middle Gila rivers that rivaled the complexity of those used in the ancient Near East, Egypt, and China. These were constructed using relatively simple excavation tools, without the benefit of advanced engineering technologies, and achieved drops of a few feet per mile, balancing erosion and siltation. The Hohokam cultivated varieties of cotton, tobacco, maize, beans and squash, as well as harvested an assortment of wild plants. Late in the Hohokam Chronological Sequence, they also used extensive dry-farming systems, primarily to grow agave for food and fiber. Their reliance on agricultural strategies based on canal irrigation, vital in their less than hospitable desert environment and arid climate, provided the basis for the aggregation of rural populations into stable urban centers.

Present Extent

Irrigation ditch in Montour County, Pennsylvania, off Strawberry Ridge Road

In the mid-20th century, the advent of diesel and electric motors led to systems that could pump groundwater out of major aquifers faster than drainage basins could refill them. This can lead to permanent loss of aquifer capacity, decreased water quality, ground subsidence, and other problems. The future of food production in such areas as the North China Plain, the Punjab, and the Great Plains of the US is threatened by this phenomenon.

At the global scale, 2,788,000 km² (689 million acres) of fertile land was equipped with irrigation infrastructure around the year 2000. About 68% of the area equipped for irrigation is located in Asia, 17% in the America, 9% in Europe, 5% in Africa and 1% in Oceania. The largest contiguous areas of high irrigation density are found:

- In Northern India and Pakistan along the Ganges and Indus rivers

- In the Hai He, Huang He and Yangtze basins in China

- Along the Nile river in Egypt and Sudan

- In the Mississippi-Missouri river basin and in parts of California

Smaller irrigation areas are spread across almost all populated parts of the world.

Only eight years later in 2008, the scale of irrigated land increased to an estimated total of 3,245,566 km² (802 million acres), which is nearly the size of India.

Types of Irrigation

Basin flood irrigation of wheat

Irrigation of land in Punjab, Pakistan

Various types of irrigation techniques differ in how the water obtained from the source is distributed within the field. In general, the goal is to supply the entire field uniformly with water, so that each plant has the amount of water it needs, neither too much nor too little.

Surface Irrigation

In *surface* (*furrow, flood,* or *level basin*) irrigation systems, water moves across the surface of agricultural lands, in order to wet it and infiltrate into the soil. Surface irrigation can be subdivided into furrow, *borderstrip or basin irrigation*. It is often called *flood irrigation* when the irrigation results in flooding or near flooding of the cultivated land. Historically, this has been the most common method of irrigating agricultural land and still used in most parts of the world.

Where water levels from the irrigation source permit, the levels are controlled by dikes, usually plugged by soil. This is often seen in terraced rice fields (rice paddies), where the method is used to flood or control the level of water in each distinct field. In some cases, the water is pumped, or lifted by human or animal power to the level of the land. The field water efficiency of surface irrigation is typically lower than other forms of irrigation but has the potential for efficiencies in the range of 70% - 90% under appropriate management.

Localized Irrigation

Impact sprinkler head

Localized irrigation is a system where water is distributed under low pressure through a piped network, in a pre-determined pattern, and applied as a small discharge to each plant or adjacent to it. Drip irrigation, spray or micro-sprinkler irrigation and bubbler irrigation belong to this category of irrigation methods.

Subsurface Textile Irrigation

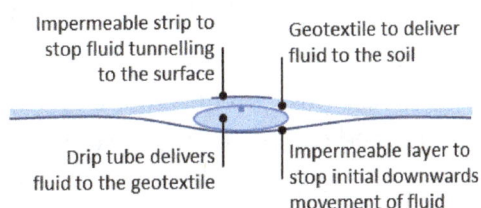

Diagram showing the structure of an example SSTI installation

Subsurface Textile Irrigation (SSTI) is a technology designed specifically for subsurface irrigation in all soil textures from desert sands to heavy clays. A typical subsurface textile irrigation system has an impermeable base layer (usually polyethylene or polypropylene), a drip line running along that base, a layer of geotextile on top of the drip line and, finally, a narrow impermeable layer on top of the geotextile. Unlike standard drip irrigation, the spacing of emitters in the drip pipe is not critical as the geotextile moves the water along the fabric up to 2 m from the dripper.

Drip Irrigation

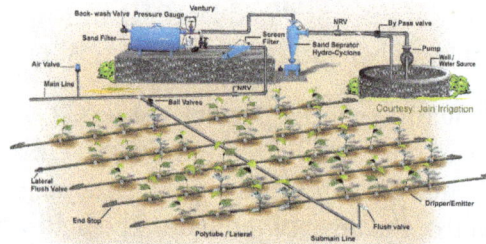

Drip irrigation layout and its parts

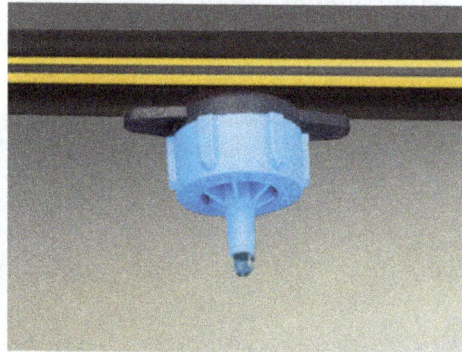

Drip irrigation – a dripper in action

Grapes in Petrolina, only made possible in this semi arid area by drip irrigation

Drip (or micro) irrigation, also known as trickle irrigation, functions as its name suggests. In this system water falls drop by drop just at the position of roots. Water is delivered at or near the root zone of plants, drop by drop. This method can be the most water-efficient method of irrigation, if managed properly, since evaporation and runoff are minimized. The field water efficiency of drip irrigation is typically in the range of 80 to 90 percent when managed correctly.

In modern agriculture, drip irrigation is often combined with plastic mulch, further reducing evaporation, and is also the means of delivery of fertilizer. The process is known as *fertigation*.

Deep percolation, where water moves below the root zone, can occur if a drip system is operated for too long or if the delivery rate is too high. Drip irrigation methods range from very high-tech and computerized to low-tech and labor-intensive. Lower water pressures are usually needed than for most other types of systems, with the exception of low energy center pivot systems and surface irrigation systems, and the system can be designed for uniformity throughout a field or for precise water delivery to individual plants in a landscape containing a mix of plant species. Although it is difficult to regulate pressure on steep slopes, pressure compensating emitters are available, so the field does not have to be level. High-tech solutions involve precisely calibrated emitters located along lines of tubing that extend from a computerized set of valves.

Irrigation Using Sprinkler Systems

Sprinkler irrigation of blueberries in Plainville, New York, United States

A traveling sprinkler at Millets Farm Centre, Oxfordshire, United Kingdom

In *sprinkler* or overhead irrigation, water is piped to one or more central locations within the field and distributed by overhead high-pressure sprinklers or guns. A system utilizing sprinklers, sprays, or guns mounted overhead on permanently installed risers is often referred to as a *solid-set* irrigation system. Higher pressure sprinklers that rotate are called *rotors* an are driven by a ball drive, gear drive, or impact mechanism. Rotors can be designed to rotate in a full or partial circle. Guns are similar to rotors, except that they generally operate at very high pressures of 40 to 130 lbf/in² (275 to 900 kPa) and flows of 50 to 1200 US gal/min (3 to 76 L/s), usually with nozzle diameters in the range of 0.5 to 1.9 inches (10 to 50 mm). Guns are used not only for irrigation, but also for industrial applications such as dust suppression and logging.

Sprinklers can also be mounted on moving platforms connected to the water source by a hose.

Automatically moving wheeled systems known as *traveling sprinklers* may irrigate areas such as small farms, sports fields, parks, pastures, and cemeteries unattended. Most of these utilize a length of polyethylene tubing wound on a steel drum. As the tubing is wound on the drum powered by the irrigation water or a small gas engine, the sprinkler is pulled across the field. When the sprinkler arrives back at the reel the system shuts off. This type of system is known to most people as a "waterreel" traveling irrigation sprinkler and they are used extensively for dust suppression, irrigation, and land application of waste water.

Other travelers use a flat rubber hose that is dragged along behind while the sprinkler platform is pulled by a cable. These cable-type travelers are definitely old technology and their use is limited in today's modern irrigation projects.

Irrigation Using Center Pivot

A small center pivot system from beginning to end

The hub of a center-pivot irrigation system

Rotator style pivot applicator sprinkler

Center pivot with drop sprinklers

Center pivot irrigation is a form of sprinkler irrigation consisting of several segments of pipe (usually galvanized steel or aluminium) joined together and supported by trusses, mounted on wheeled towers with sprinklers positioned along its length. The system moves in a circular pattern and is fed with water from the pivot point at the center of the arc. These systems are found and used in all parts of the world and allow irrigation of all types of terrain. Newer systems have drop sprinkler heads as shown in the image that follows.

Wheel line irrigation system in Idaho, 2001

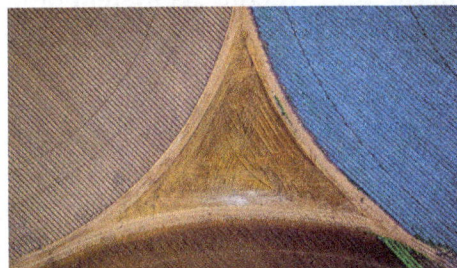

Center pivot irrigation

Most center pivot systems now have drops hanging from a u-shaped pipe attached at the top of the pipe with sprinkler head that are positioned a few feet (at most) above the crop, thus limiting evaporative losses. Drops can also be used with drag hoses or bubblers that deposit the water directly on the ground between crops. Crops are often planted in a circle to conform to the center pivot. This type of system is known as LEPA (Low Energy Precision Application). Originally, most center pivots were water powered. These were replaced by hydraulic systems (*T-L Irrigation*) and electric motor driven systems (Reinke, Valley, Zimmatic). Many modern pivots feature GPS devices.

Irrigation by Lateral Move (Side Roll, Wheel Line, Wheelmove)

A *series of pipes, each with a wheel* of about 1.5 m diameter permanently affixed to its mid-point, and sprinklers along its length, are coupled together. Water is supplied at one end using a large hose. After sufficient irrigation has been applied to one strip of the field, the hose is removed, the water drained from the system, and the assembly rolled either by hand or with a purpose-built mechanism, so that the sprinklers are moved to a different position across the field. The hose is reconnected. The process is repeated in a pattern until the whole field has been irrigated.

This system is less expensive to install than a center pivot, but much more labor-intensive to operate - it does not travel automatically across the field: it applies water in a stationary strip, must be drained, and then rolled to a new strip. Most systems use 4 or 5-inch (130 mm) diameter aluminum pipe. The pipe doubles both as water transport and as an axle for rotating all the wheels. A drive system (often found near the centre of the wheel line) rotates the clamped-together pipe sections as a single axle, rolling the whole wheel line. Manual adjustment of individual wheel positions may be necessary if the system becomes misaligned.

Wheel line systems are limited in the amount of water they can carry, and limited in the height of crops that can be irrigated. One useful feature of a lateral move system is that it consists of sections that can be easily disconnected, adapting to field shape as the line is moved. They are most often used for small, rectilinear, or oddly-shaped fields, hilly or mountainous regions, or in regions where labor is inexpensive.

Sub-irrigation

Subirrigation has been used for many years in field crops in areas with high water tables. It is a method of artificially raising the water table to allow the soil to be moistened from below the plants' root zone. Often those systems are located on permanent grasslands in lowlands or river valleys and combined with drainage infrastructure. A system of pumping stations, canals, weirs and gates allows it to increase or decrease the water level in a network of ditches and thereby control the water table.

Sub-irrigation is also used in commercial greenhouse production, usually for potted plants. Water is delivered from below, absorbed upwards, and the excess collected for recycling. Typically, a solution of water and nutrients floods a container or flows through a trough for a short period of time, 10–20 minutes, and is then pumped back into a holding tank for reuse. Sub-irrigation in greenhouses requires fairly sophisticated, expensive equipment and management. Advantages are water and nutrient conservation, and labor-saving through lowered system maintenance and automation. It is similar in principle and action to subsurface basin irrigation.

Irrigation Automatically, Non-electric Using Buckets and Ropes

Besides the common manual watering by bucket, an automated, natural version of this also exists. Using plain polyester ropes combined with a prepared ground mixture can be used to water plants from a vessel filled with water.

The ground mixture would need to be made depending on the plant itself, yet would mostly consist of black potting soil, vermiculite and perlite. This system would (with certain crops) allow to save expenses as it does not consume any electricity and only little water (unlike sprinklers, water timers, etc.). However, it may only be used with certain crops (probably mostly larger crops that do not need a humid environment; perhaps e.g. paprikas).

Irrigation Using Water Condensed from Humid Air

In countries where at night, humid air sweeps the countryside.Water can be obtained from the humid air by condensation onto cold surfaces. This is for example practiced in the vineyards at Lanzarote using stones to condense water or with various fog collectors based on canvas or foil sheets.

In-ground Irrigation

Most commercial and residential irrigation systems are "in ground" systems, which means that everything is buried in the ground. With the pipes, sprinklers, emitters (drippers), and irrigation valves being hidden, it makes for a cleaner, more presentable landscape without garden hoses or other items having to be moved around manually. This does, however, create some drawbacks in the maintenance of a completely buried system.

Most irrigation systems are divided into zones. A zone is a single irrigation valve and one or a group of drippers or sprinklers that are connected by pipes or tubes. Irrigation systems are divided into zones because there is usually not enough pressure and available flow to run sprinklers for an entire yard or sports field at once. Each zone has a solenoid valve on it that is controlled via wire by an irrigation controller. The irrigation controller is either a mechanical (now the "dinosaur" type) or electrical device that signals a zone to turn on at a specific time and keeps it on for a specified amount of time. "Smart Controller" is a recent term for a controller that is capable of adjusting the watering time by itself in response to current environmental conditions. The smart controller determines current conditions by means of historic weather data for the local area, a soil moisture sensor (water potential or water content), rain sensor, or in more sophisticated systems satellite feed weather station, or a combination of these.

When a zone comes on, the water flows through the lateral lines and ultimately ends up at the irrigation emitter (drip) or sprinkler heads. Many sprinklers have pipe thread inlets on the bottom of them which allows a fitting and the pipe to be attached to them. The sprinklers are usually installed with the top of the head flush with the ground surface. When the water is pressurized, the head will pop up out of the ground and water the desired area until the valve closes and shuts off that zone. Once there is no more water pressure in the lateral line, the sprinkler head will retract back into the ground. Emitters are generally laid on the soil surface or buried a few inches to reduce evaporation losses.

Water Sources

Irrigation water can come from groundwater (extracted from springs or by using wells), from surface water (withdrawn from rivers, lakes or reservoirs) or from non-conventional sources like treated wastewater, desalinated water or drainage water. A special form of irrigation using surface water is spate irrigation, also called floodwater harvesting. In case of a flood (spate), water is diverted to normally dry river beds (wadis) using a network of dams, gates and channels and spread over large areas. The moisture stored in the soil will be used thereafter to grow crops. Spate irrigation areas are in particular located in semi-arid or arid, mountainous regions. While flood-water harvesting belongs to the accepted irrigation methods, rainwater harvesting is usually not considered as a form of irrigation. Rainwater harvesting is the collection of runoff water from roofs or unused land and the concentration of this.

Irrigation is underway by pump-enabled extraction directly from the Gumti, seen in the background, in Comilla, Bangladesh.

Around 90% of wastewater produced globally remains untreated, causing widespread water pollution, especially in low-income countries. Increasingly, agriculture uses untreated wastewater as a source of irrigation water. Cities provide lucrative markets for fresh produce, so are attractive to farmers. However, because agriculture has to compete for increasingly scarce water resources with industry and municipal users, there is often no alternative for farmers but to use water polluted with urban waste, including sewage, directly to water their crops. Significant health hazards can result from using water loaded with pathogens in this way, especially if people eat raw vegetables that have been irrigated with the polluted water. The International Water Management Institute has worked in India, Pakistan, Vietnam, Ghana, Ethiopia, Mexico and other countries on various projects aimed at assessing and reducing risks of wastewater irrigation. They advocate a 'multiple-barrier' approach to wastewater use, where farmers are encouraged to adopt various risk-reducing behaviours. These include ceasing irrigation a few days before harvesting to allow pathogens to die off in the sunlight, applying water carefully so it does not contaminate leaves likely to be eaten raw, cleaning vegetables with disinfectant or allowing fecal sludge used in farming to dry before being used as a human manure. The World Health Organization has developed guidelines for safe water use.

There are numerous benefits of using recycled water for irrigation, including the low cost (when compared to other sources, particularly in an urban area), consistency of supply (regardless of season, climatic conditions and associated water restrictions), and general consistency of quality. Irrigation of recycled wastewater is also considered as a means for plant fertilization and particularly nutrient supplementation. This approach carries with it a risk of soil and water pollution through excessive wastewater application. Hence, a detailed understanding of soil water conditions is essential for effective utilization of wastewater for irrigation.

Efficiency

Young engineers restoring and developing the old Mughal irrigation system during the reign of the
Mughal Emperor Bahadur Shah II

Modern irrigation methods are efficient enough to supply the entire field uniformly with water, so that each plant has the amount of water it needs, neither too much nor too little. Water use efficiency in the field can be determined as follows:

- Field Water Efficiency (%) = (Water Transpired by Crop ÷ Water Applied to Field) x 100

Until 1960s, the common perception was that water was an infinite resource. At that time, there were fewer than half the current number of people on the planet. People were not as wealthy as today, consumed fewer calories and ate less meat, so less water was needed to produce their food. They required a third of the volume of water we presently take from rivers. Today, the competition for water resources is much more intense. This is because there are now more than seven billion people on the planet, their consumption of water-thirsty meat and vegetables is rising, and there is increasing competition for water from industry, urbanisation and biofuel crops. To avoid a global water crisis, farmers will have to strive to increase productivity to meet growing demands for food, while industry and cities find ways to use water more efficiently.

Successful agriculture is dependent upon farmers having sufficient access to water. However, water scarcity is already a critical constraint to farming in many parts of the world. With regards to agriculture, the World Bank targets food production and water management as an increasingly global issue that is fostering a growing debate. Physical water scarcity is where there is not enough water to meet all demands, including that needed for ecosystems to function effectively. Arid regions frequently suffer from physical water scarcity. It also occurs where water seems abundant but where resources are over-committed. This can happen where there is overdevelopment of hydraulic infrastructure, usually for irrigation. Symptoms of physical water scarcity include environmental degradation and declining groundwater. Economic scarcity, meanwhile, is caused by a lack of investment in water or insufficient human capacity to satisfy the demand for water. Symptoms of economic water scarcity include a lack of infrastructure, with people often having to fetch water from rivers for domestic and agricultural uses. Some 2.8 billion people currently live in water-scarce areas.

Technical Challenges

Irrigation schemes involve solving numerous engineering and economic problems while minimizing negative environmental impact.

- Competition for surface water rights.

- Overdrafting (depletion) of underground aquifers.

- Ground subsidence (e.g. New Orleans, Louisiana)

- Underirrigation or irrigation giving only just enough water for the plant (e.g. in drip line irrigation) gives poor soil salinity control which leads to increased soil salinity with consequent buildup of toxic salts on soil surface in areas with high evaporation. This requires either leaching to remove these salts and a method of drainage to carry the salts away. When using drip lines, the leaching is best done regularly at certain intervals (with only a slight excess of water), so that the salt is flushed back under the plant's roots.

- Overirrigation because of poor distribution uniformity or management wastes water, chemicals, and may lead to water pollution.

- Deep drainage (from over-irrigation) may result in rising water tables which in some instances will lead to problems of irrigation salinity requiring watertable control by some form of subsurface land drainage.

- Irrigation with saline or high-sodium water may damage soil structure owing to the formation of alkaline soil

- Clogging of filters: It is mostly algae that clog filters, drip installations and nozzles. UV and ultrasonic method can be used for algae control in irrigation systems.

References

- H.A. Mills; J.B. Jones Jr. (1996). Plant Analysis Handbook II: A practical Sampling, Preparation, Analysis, and Interpretation Guide. ISBN 1-878148-05-2.

- J. Benton Jones, Jr. "Inorganic Chemical Fertilizers and Their Properties" in Plant Nutrition and Soil Fertility Manual, Second Edition. CRC Press, 2012. ISBN 978-1-4398-1609-7. eBook ISBN 978-1-4398-1610-3.

- Moore, Geoff (2001). Soilguide - A handbook for understanding and managing agricultural soils (PDF). Perth, Western Australia: Agriculture Western Australia. pp. 161–207. ISBN 0 7307 0057 7.

- Appl, Max (2000). Ullmann's Encyclopedia of Industrial Chemistry, Volume 3. Weinheim, Germany: Wiley-VCH Verlag GmbH & Co. KGaA. pp. 139–225. doi:10.1002/14356007.002_011. ISBN 9783527306732.

- G. J. Leigh (2004). The world's greatest fix: a history of nitrogen and agriculture. Oxford University Press US. pp. 134–139. ISBN 0-19-516582-9.

- Trevor Illtyd Williams; Thomas Kingston Derry (1982). A short history of twentieth-century technology c. 1900-c. 1950. Oxford University Press. pp. 134–135. ISBN 0-19-858159-9.

- Francis Borgio J, Sahayaraj K and Alper Susurluk I (eds) . Microbial Insecticides: Principles and Applications, Nova Publishers, USA. 492pp. ISBN 978-1-61209-223-2

- Quastel, J. H. (1950). "2,4-Dichlorophenoxyacetic Acid (2,4-D) as a Selective Herbicide". Agricultural Control Chemicals. Advances in Chemistry. 1. p. 244. doi:10.1021/ba-1950-0001.ch045. ISBN 0-8412-2442-0.

- Smith (18 July 1995). "8: Fate of herbicides in the environment". Handbook of Weed Management Systems. CRC Press. pp. 245–278. ISBN 978-0-8247-9547-4.

6

Environmental Impacts of Agriculture

The environmental impact of agriculture depends on the practices that are performed by the farmers of that particular region. The aspects discussed in this chapter are climate change and agriculture, genetically modified crops, environmental impact of irrigation, environmental impact of pesticides and plasticulture. The section serves as a source to understand the major impacts of agriculture.

Environmental Impact of Agriculture

Water pollution in a rural stream due to runoff from farming activity in New Zealand.

The environmental impact of agriculture varies based on the wide variety of agricultural practices employed around the world. Ultimately, the environmental impact depends on the production practices of the system used by farmers. The connection between emissions into the environment and the farming system is indirect, as it also depends on other climate variables such as rainfall and temperature.

There are two types of indicators of environmental impact: "means-based", which is based on the farmer's production methods, and "effect-based", which is the impact that farming methods have on the farming system or on emissions to the environment. An example of a means-based indicator would be the quality of groundwater, that is effected by the amount of nitrogen applied to the soil. An indicator reflecting the loss of nitrate to groundwater would be effect-based.

The environmental impact of agriculture involves a variety of factors from the soil, to water, the air, animal and soil diversity, people, plants, and the food itself. Some of the environmental issues that are related to agriculture are climate change, deforestation, genetic engineering, irrigation problems, pollutants, soil degradation, and waste.

Issues

Climate Change

Climate change and agriculture are interrelated processes, both of which take place on a worldwide scale. Global warming is projected to have significant impacts on conditions affecting agriculture, including temperature, precipitation and glacial run-off. These conditions determine the carrying capacity of the biosphere to produce enough food for the human population and domesticated animals. Rising carbon dioxide levels would also have effects, both detrimental and beneficial, on crop yields. Assessment of the effects of global climate changes on agriculture might help to properly anticipate and adapt farming to maximize agricultural production. Although the net impact of climate change on agricultural production is uncertain it is likely that it will shift the suitable growing zones for individual crops. Adjustment to this geographical shift will involve considerable economic costs and social impacts.

At the same time, agriculture has been shown to produce significant effects on climate change, primarily through the production and release of greenhouse gases such as carbon dioxide, methane, and nitrous oxide. In addition, agriculture that practices tillage, fertilization, and pesticide application also releases ammonia, nitrate, phosphorus, and many other pesticides that affect air, water, and soil quality, as well as biodiversity. Agriculture also alters the Earth's land cover, which can change its ability to absorb or reflect heat and light, thus contributing to radiative forcing. Land use change such as deforestation and desertification, together with use of fossil fuels, are the major anthropogenic sources of carbon dioxide; agriculture itself is the major contributor to increasing methane and nitrous oxide concentrations in earth's atmosphere.

Deforestation

Deforestation is clearing the Earth's forests on a large scale worldwide and resulting in many land damages. One of the causes of deforestation is to clear land for pasture or crops. According to British environmentalist Norman Myers, 5% of deforestation is due to cattle ranching, 19% due to over-heavy logging, 22% due to the growing sector of palm oil plantations, and 54% due to slash-and-burn farming.

Deforestation causes the loss of habitat for millions of species, and is also a driver of climate change. Trees act as a carbon sink: that is, they absorb carbon dioxide, an unwanted greenhouse gas, out of the atmosphere. Removing trees releases carbon dioxide into the atmosphere and leaves behind fewer trees to absorb the increasing amount of carbon dioxide in the air. In this way, deforestation exacerbates climate change. When trees are removed from forests, the soils tend to dry out because there is no longer shade, and there are not enough trees to assist in the water cycle by returning water vapor back to the environment. With no trees, landscapes that were once forests can potentially become barren deserts. The removal of trees also causes extreme fluctuations in temperature.

In 2000 the United Nations Food and Agriculture Organization (FAO) found that "the role of population dynamics in a local setting may vary from decisive to negligible," and that deforestation can result from "a combination of population pressure and stagnating economic, social and technological conditions."

Genetic Engineering

Genetically engineered crops are herbicide-tolerant, and their overuse has created herbicide resistant "super weeds",which may ultimately increase the use of herbicides. Seed contamination is another problem of genetic engineering; it can occur from wind or bee pollination that is blown from genetically-engineered crops to normal crops. About 50% of corn and soybean samples and more than 80% of canola samples were found to be contaminated by Monsanto's (genetic engineering company) genes.This accidental contamination can cause organic farmers to lose a lot of money because they need to recall their products. There are various cases of this such as in the corn and alfalfa industry.

Irrigation

Irrigation can lead to a number of problems:

Among some of these problems is the depletion of underground aquifers through overdrafting. Soil can be over-irrigated because of poor distribution uniformity or management wastes water, chemicals, and may lead to water pollution. Over-irrigation can cause deep drainage from rising water tables that can lead to problems of irrigation salinity requiring watertable control by some form of subsurface land drainage. However, if the soil is under irrigated, it gives poor soil salinity control which leads to increased soil salinity with consequent buildup of toxic salts on soil surface in areas with high evaporation. This requires either leaching to remove these salts and a method of drainage to carry the salts away. Irrigation with saline or high-sodium water may damage soil structure owing to the formation of alkaline soil.

Pollutants

Synthetic pesticides such as 'Malathon, 'Rogor', 'Kelthane' and 'confidor' are the most widespread method of controlling pests in agriculture. Pesticides can leach through the soil and enter the groundwater, as well as linger in food products and result in death in humans. Pesticides can also kill non-target plants, birds, fish and other wildlife. A wide range of agricultural chemicals are used and some become pollutants through use, misuse, or ignorance. Pollutants from agriculture have a huge effect on water quality. Agricultural nonpoint source (NPS) solution impacts lakes, rivers, wetlands, estuaries, and groundwater. Agricultural NPS can be caused by poorly managed animal feeding operations, overgrazing, plowing, fertilizer, and improper, excessive, or badly timed use of Pesticides. Pollutants from farming include sediments, nutrients, pathogens, pesticides, metals, and salts.

Listed below are additional and specific problems that may arise with the release of pollutants from agriculture.

- Pesticide drift

o   soil contamination

o   air pollution *spray drift*

- Pesticides, especially those based on organochloride

- Pesticide residue in foods

- Pesticide toxicity to bees

  o   List of crop plants pollinated by bees

  o   Pollination management

- Bioremediation

Soil Degradation

Soil degradation is the decline in soil quality that can be a result of many factors, especially from agriculture. Soils hold the majority of the world's biodiversity, and healthy soils are essential for food production and an adequate water supply. Common attributes of soil degradation can be salting, waterlogging, compaction, pesticide contamination, decline in soil structure quality, loss of fertility, changes in soil acidity, alkalinity, salinity, and erosion. Soil degradation also has a huge impact on biological degradation, which affects the microbial community of the soil and can alter nutrient cycling, pest and disease control, and chemical transformation properties of the soil.

- soil contamination

  o   sedimentation

Waste

Plasticulture is the use of plastic mulch in agriculture. Farmers use plastic sheets as mulch to cover 50-70% of the soil and allows them to use drip irrigation systems to have better control over soil nutrients and moisture. Rain is not required in this system, and farms that use plasticulture are built to encourage the fastest runoff of rain. The use of pesticides with plasticulture allows pesticides to be transported easier in the surface runoff towards wetlands or tidal creeks. The runoff from pesticides and chemicals in the plastic can cause serious deformations and death in shellfish as the runoff carries the chemicals towards the oceans.

In addition to the increased runoff that results from plasticulture, there is also the problem of the increased amount of waste form the plastic mulch itself. The use of plastic mulch for vegetables, strawberries, and other row and orchard crops exceeds 110 million pounds annually in the United States. Most plastic ends up in the landfill, although there are other disposal options such as disking mulches into the soil, on-site burying, on-site storage, reuse, recycling, and incineration. The incineration and recycling options are complicated by the variety of the types of plastics that are used and by the geographic dispersal of the plastics. Plastics also contain stabilizers and dyes as well as heavy metals, which limits the amount of products that can be recycled. Research is continually being conducted on creating biodegradable or photodegradable mulches. While there has been minor success with this, there is also the problem of how long the plastic takes to degrade, as many biodegradable products take a long time to break down.

Issues by Region

The environmental impact of agriculture can vary depending on the region as well as the type of agriculture production method that is being used. Listed below are some specific environmental issues in a various different regions around the world.

- Hedgerow removal in the United Kingdom.

- Soil salinisation, especially in Australia.

- Phosphate mining in Nauru

- Methane emissions from livestock in New Zealand.

- Environmentalists attribute the hypoxic zone in the Gulf of Mexico as being encouraged by nitrogen fertilization of the algae bloom.

Sustainable Agriculture

The exponential population increase in recent decades has increased the practice of agricultural land conversion to meet demand for food which in turn has increased the effects on the environment. The global population is still increasing and will eventually stabilise, as some critics doubt that food production, due to lower yields from global warming, can support the global population.

Organic farming is a multifaceted sustainable agriculture set of practices that can have a lower impact on the environment at the small scale. However in most cases organic farming results in lower yields in terms of production per unit area and per unit of irrigation water. Therefore, widespread adoption of organic agriculture will require additional land to be cleared and water resources extracted to meet the same level of production. A European meta-analysis found that organic farms tended to have higher soil organic matter content and lower nutrient losses (nitrogen leaching, nitrous oxide emissions and ammonia emissions) per unit of field area but higher ammonia emissions, nitrogen leaching and nitrous oxide emissions per product unit.

Other specific methods include: permaculture; and biodynamic agriculture which incorporates a spiritual element.

- Category: Sustainable agriculture

- Biological pest control

Climate Change and Agriculture

Climate change and agriculture are interrelated processes, both of which take place on a global scale. Climate change affects agriculture in a number of ways, including through changes in average temperatures, rainfall, and climate extremes (e.g., heat waves); changes in pests and diseases; changes in atmospheric carbon dioxide and ground-level ozone concentrations; changes in the nutritional quality of some foods; and changes in sea level.

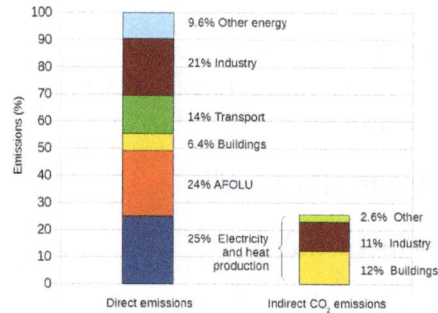

Human greenhouse gas emissions by sector, in the year 2010. "AFOLU" stands for "agriculture, forestry, and other land use".

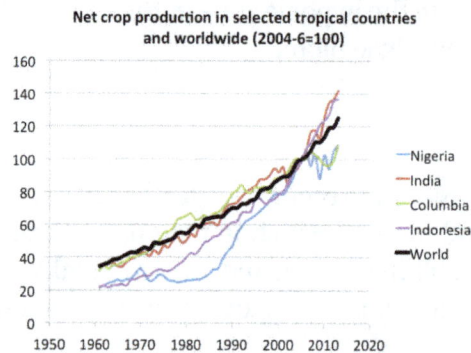

Graph of net crop production worldwide and in selected tropical countries. Raw data from the United Nations.

Climate change is already affecting agriculture, with effects unevenly distributed across the world. Future climate change will likely negatively affect crop production in low latitude countries, while effects in northern latitudes may be positive or negative. Climate change will probably increase the risk of food insecurity for some vulnerable groups, such as the poor.

Agriculture contributes to climate change by (1) anthropogenic emissions of greenhouse gases (GHGs), and (2) by the conversion of non-agricultural land (e.g., forests) into agricultural land. Agriculture, forestry and land-use change contributed around 20 to 25% to global annual emissions in 2010.

There are range of policies that can reduce the risk of negative climate change impacts on agriculture, and to reduce GHG emissions from the agriculture sector.

Impact of Climate Change on Agriculture

Despite technological advances, such as improved varieties, genetically modified organisms, and irrigation systems, weather is still a key factor in agricultural productivity, as well as soil properties and natural communities. The effect of climate on agriculture is related to variabilities in local climates rather than in global climate patterns. The Earth's average surface temperature has increased by 1.5 °F (0.83 °C) since 1880. Consequently, agronomists consider any assessment has to be individually consider each local area.

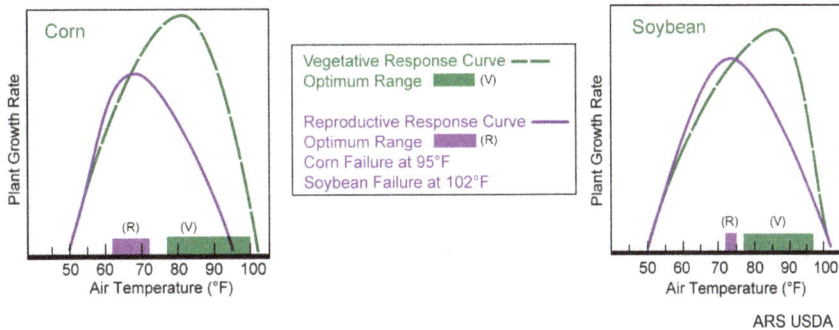

For each plant variety, there is an optimal temperature for vegetative growth, with growth dropping off as temperatures increase or decrease. Similarly, there is a range of temperatures at which a plant will produce seed. Outside of this range, the plant will not reproduce. As the graphs show, corn will fail to reproduce at temperatures above 95 °F and soybean above 102 °F.

On the other hand, agricultural trade has grown in recent years, and now provides significant amounts of food, on a national level to major importing countries, as well as comfortable income to exporting ones. The international aspect of trade and security in terms of food implies the need to also consider the effects of climate change on a global scale.

A 2008 study published in *Science* suggested that, due to climate change, "southern Africa could lose more than 30% of its main crop, maize, by 2030. In South Asia losses of many regional staples, such as rice, millet and maize could top 10%".

The Intergovernmental Panel on Climate Change (IPCC) has produced several reports that have assessed the scientific literature on climate change. The IPCC Third Assessment Report, published in 2001, concluded that the poorest countries would be hardest hit, with reductions in crop yields in most tropical and sub-tropical regions due to decreased water availability, and new or changed insect pest incidence. In Africa and Latin America many rainfed crops are near their maximum temperature tolerance, so that yields are likely to fall sharply for even small climate changes; falls in agricultural productivity of up to 30% over the 21st century are projected. Marine life and the fishing industry will also be severely affected in some places.

Climate change induced by increasing greenhouse gases is likely to affect crops differently from region to region. For example, average crop yield is expected to drop down to 50% in Pakistan according to the UKMO scenario whereas corn production in Europe is expected to grow up to 25% in optimum hydrologic conditions.

More favourable effects on yield tend to depend to a large extent on realization of the potentially beneficial effects of carbon dioxide on crop growth and increase of efficiency in water use. Decrease in potential yields is likely to be caused by shortening of the growing period, decrease in water availability and poor vernalization.

In the long run, the climatic change could affect agriculture in several ways :

- *productivity*, in terms of quantity and quality of crops

- *agricultural practices*, through changes of water use (irrigation) and agricultural inputs such as herbicides, insecticides and fertilizers

- *environmental effects*, in particular in relation of frequency and intensity of soil drainage (leading to nitrogen leaching), soil erosion, reduction of crop diversity

- *rural space*, through the loss and gain of cultivated lands, land speculation, land renunciation, and hydraulic amenities.

- *adaptation*, organisms may become more or less competitive, as well as humans may develop urgency to develop more competitive organisms, such as flood resistant or salt resistant varieties of rice.

They are large uncertainties to uncover, particularly because there is lack of information on many specific local regions, and include the uncertainties on magnitude of climate change, the effects of technological changes on productivity, global food demands, and the numerous possibilities of adaptation.

Most agronomists believe that agricultural production will be mostly affected by the severity and pace of climate change, not so much by gradual trends in climate. If change is gradual, there may be enough time for biota adjustment. Rapid climate change, however, could harm agriculture in many countries, especially those that are already suffering from rather poor soil and climate conditions, because there is less time for optimum natural selection and adaption.

But much remains unknown about exactly how climate change may affect farming and food security, in part because the role of farmer behaviour is poorly captured by crop-climate models. For instance, Evan Fraser, a geographer at the University of Guelph in Ontario Canada, has conducted a number of studies that show that the socio-economic context of farming may play a huge role in determining whether a drought has a major, or an insignificant impact on crop production. In some cases, it seems that even minor droughts have big impacts on food security (such as what happened in Ethiopia in the early 1980s where a minor drought triggered a massive famine), versus cases where even relatively large weather-related problems were adapted to without much hardship. Evan Fraser combines socio-economic models along with climatic models to identify "vulnerability hotspots" One such study has identified US maize (corn) production as particularly vulnerable to climate change because it is expected to be exposed to worse droughts, but it does not have the socio-economic conditions that suggest farmers will adapt to these changing conditions.

Observed Impacts

So far, the effects of regional climate change on agriculture have been relatively limited. Changes in crop phenology provide important evidence of the response to recent regional climate change. Phenology is the study of natural phenomena that recur periodically, and how these phenomena relate to climate and seasonal changes. A significant advance in phenology has been observed for agriculture and forestry in large parts of the Northern Hemisphere.

Droughts have been occurring more frequently because of global warming and they are expected to become more frequent and intense in Africa, southern Europe, the Middle East, most of the Americas, Australia, and Southeast Asia. Their impacts are aggravated because of increased water demand, population growth, urban expansion, and environmental protection efforts in many areas. Droughts result in crop failures and the loss of pasture grazing land for livestock.

Examples

As of the decade starting in 2010, many hot countries have thriving agricultural sectors.

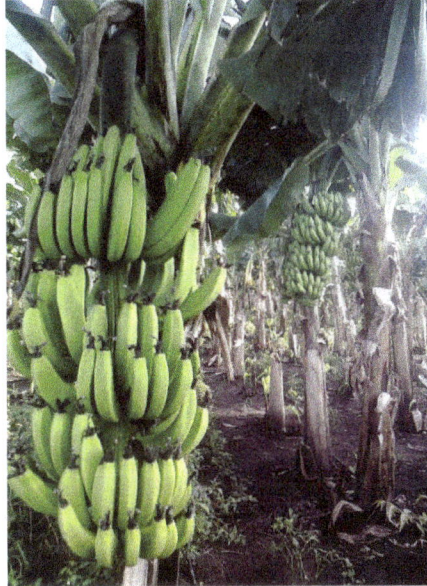

Banana farm at Chinawal village in Jalgaon district, India

Jalgaon district, India, has an average temperature which ranges from 20.2C in December to 29.8C in May, and an average precipitation of 750mm/year. It produces bananas at a rate that would make it the world's seventh-largest banana producer if it were a country.

During the period 1990-2012, Nigeria had an average temperature which ranged from a low of 24.9C in January to a high of 30.4C in April. According to the Food and Agriculture Organization of the United Nations (FAO), Nigeria is by far the world's largest producer of yams, producing over 38 million tonnes in 2012. The second through 8th largest yam producers were all nearby African countries, with the largest non-African producer, Papua New Guinea, producing less than 1% of Nigerian production.

In 2013, according to the FAO, Brazil and India were by far the world's leading producers of Sugarcane, with a combined production of over 1 billion tonnes, or over half of worldwide production.

Projections

As part of the IPCC's Fourth Assessment Report, Schneider *et al.* (2007) projected the potential future effects of climate change on agriculture. With low to medium confidence, they concluded that for about a 1 to 3 °C global mean temperature increase (by 2100, relative to the 1990–2000 average level) there would be productivity decreases for some cereals in low latitudes, and productivity increases in high latitudes. In the IPCC Fourth Assessment Report, "low confidence" means that a particular finding has about a 2 out of 10 chance of being correct, based on expert judgement. "Medium confidence" has about a 5 out of 10 chance of being correct. Over the same time period, with medium confidence, global production potential was projected to:

- increase up to around 3 °C,

- very likely decrease above about 3 °C.

Most of the studies on global agriculture assessed by Schneider *et al.* (2007) had not incorporated a number of critical factors, including changes in extreme events, or the spread of pests and diseases. Studies had also not considered the development of specific practices or technologies to aid adaptation to climate change.

The US National Research Council (US NRC, 2011) assessed the literature on the effects of climate change on crop yields. US NRC (2011) stressed the uncertainties in their projections of changes in crop yields.

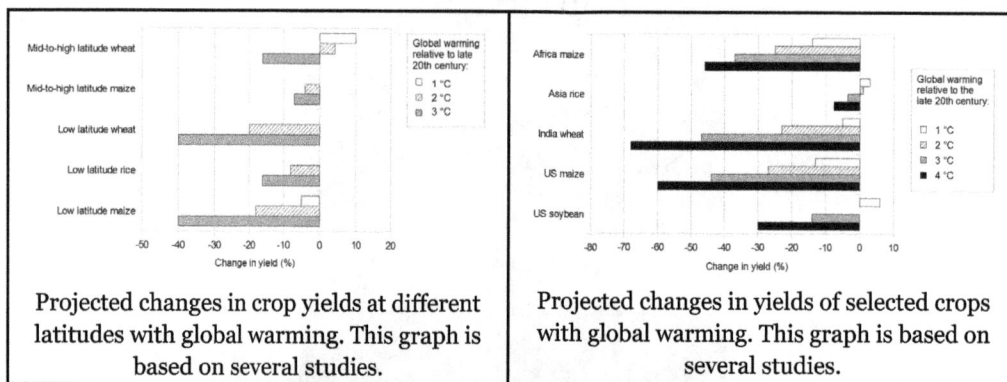

| Projected changes in crop yields at different latitudes with global warming. This graph is based on several studies. | Projected changes in yields of selected crops with global warming. This graph is based on several studies. |

Their central estimates of changes in crop yields are shown above. Actual changes in yields may be above or below these central estimates. US NRC (2011) also provided an estimated the "likely" range of changes in yields. "Likely" means a greater than 67% chance of being correct, based on expert judgement. The likely ranges are summarized in the image descriptions of the two graphs.

Food Security

The IPCC Fourth Assessment Report also describes the impact of climate change on food security. Projections suggested that there could be large decreases in hunger globally by 2080, compared to the (then-current) 2006 level. Reductions in hunger were driven by projected social and economic development. For reference, the Food and Agriculture Organization has estimated that in 2006, the number of people undernourished globally was 820 million. Three scenarios *without* climate change (SRES A1, B1, B2) projected 100-130 million undernourished by the year 2080, while another scenario without climate change (SRES A2) projected 770 million undernourished. Based on an expert assessment of all of the evidence, these projections were thought to have about a 5-in-10 chance of being correct.

The same set of greenhouse gas and socio-economic scenarios were also used in projections that included the effects of climate change. *Including* climate change, three scenarios (SRES A1, B1, B2) projected 100-380 million undernourished by the year 2080, while another scenario with climate change (SRES A2) projected 740-1,300 million undernourished. These projections were thought to have between a 2-in-10 and 5-in-10 chance of being correct.

Projections also suggested regional changes in the global distribution of hunger. By 2080, sub-Saharan Africa may overtake Asia as the world's most food-insecure region. This is mainly due to projected social and economic changes, rather than climate change.

Individual Studies

Cline (2008) looked at how climate change might affect agricultural productivity in the 2080s. His study assumes that no efforts are made to reduce anthropogenic greenhouse gas emissions, leading to global warming of 3.3 °C above the pre-industrial level. He concluded that global agricultural productivity could be negatively affected by climate change, with the worst effects in developing countries.

Lobell *et al.* (2008a) assessed how climate change might affect 12 food-insecure regions in 2030. The purpose of their analysis was to assess where adaptation measures to climate change should be prioritized. They found that without sufficient adaptation measures, South Asia and South Africa would likely suffer negative impacts on several crops which are important to large food insecure human populations.

Battisti and Naylor (2009) looked at how increased seasonal temperatures might affect agricultural productivity. Projections by the IPCC suggest that with climate change, high seasonal temperatures will become widespread, with the likelihood of extreme temperatures increasing through the second half of the 21st century. Battisti and Naylor (2009) concluded that such changes could have very serious effects on agriculture, particularly in the tropics. They suggest that major, near-term, investments in adaptation measures could reduce these risks.

"Climate change merely increases the urgency of reforming trade policies to ensure that global food security needs are met" said C. Bellmann, ICTSD Programmes Director. A 2009 ICTSD-IPC study by Jodie Keane suggests that climate change could cause farm output in sub-Saharan Africa to decrease by 12 percent by 2080 - although in some African countries this figure could be as much as 60 percent, with agricultural exports declining by up to one fifth in others. Adapting to climate change could cost the agriculture sector $14bn globally a year, the study finds.

Regional

Africa

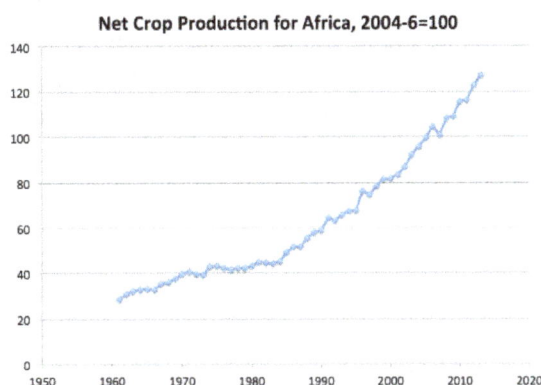

African crop production. Raw data from the United Nations.

In Africa, IPCC (2007:13) projected that climate variability and change would severely compromise agricultural production and access to food. This projection was assigned "high confidence."

Africa's geography makes it particularly vulnerable to climate change, and seventy per cent of the population rely on rain-fed agriculture for their livelihoods. Tanzania's official report on climate change suggests that the areas that usually get two rainfalls in the year will probably get more, and those that get only one rainy season will get far less. As of 2005, the net result was expected to be that 33% less maize—the country's staple crop—would be grown.

Asia

In East and Southeast Asia, IPCC (2007:13) projected that crop yields could increase up to 20% by the mid-21st century. In Central and South Asia, projections suggested that yields might decrease by up to 30%, over the same time period. These projections were assigned "medium confidence." Taken together, the risk of hunger was projected to remain very high in several developing countries.

More detailed analysis of rice yields by the International Rice Research Institute forecast 20% reduction in yields over the region per degree Celsius of temperature rise. Rice becomes sterile if exposed to temperatures above 35 degrees for more than one hour during flowering and consequently produces no grain.

A 2013 study by the International Crops Research Institute for the Semi-Arid Tropics (ICRISAT) aimed to find science-based, pro-poor approaches and techniques that would enable Asia's agricultural systems to cope with climate change, while benefitting poor and vulnerable farmers. The study's recommendations ranged from improving the use of climate information in local planning and strengthening weather-based agro-advisory services, to stimulating diversification of rural household incomes and providing incentives to farmers to adopt natural resource conservation measures to enhance forest cover, replenish groundwater and use renewable energy. A 2014 study found that warming had increased maize yields in the Heilongjiang region of China had increased by between 7 and 17% per decade as a result of rising temperatures.

Australia and New Zealand

Hennessy *et al..* (2007:509) assessed the literature for Australia and New Zealand. They concluded that without further adaptation to climate change, projected impacts would likely be substantial: By 2030, production from agriculture and forestry was projected to decline over much of southern and eastern Australia, and over parts of eastern New Zealand; In New Zealand, initial benefits were projected close to major rivers and in western and southern areas. Hennessy *et al..* (2007:509) placed high confidence in these projections.

Europe

With high confidence, IPCC (2007:14) projected that in Southern Europe, climate change would reduce crop productivity. In Central and Eastern Europe, forest productivity was expected to decline. In Northern Europe, the initial effect of climate change was projected to increase crop yields.

Latin America

The major agricultural products of Latin American regions include livestock and grains, such as maize, wheat, soybeans, and rice. Increased temperatures and altered hydrological cycles are predicted to translate to shorter growing seasons, overall reduced biomass production, and lower grain yields. Brazil, Mexico and Argentina alone contribute 70-90% of the total agricultural production in Latin America. In these and other dry regions, maize production is expected to decrease. A study summarizing a number of impact studies of climate change on agriculture in Latin America indicated that wheat is expected to decrease in Brazil, Argentina and Uruguay. Livestock, which is the main agricultural product for parts of Argentina, Uruguay, southern Brazil, Venezuela, and Colombia is likely to be reduced. Variability in the degree of production decrease among different regions of Latin America is likely. For example, one 2003 study that estimated future maize production in Latin America predicted that by 2055 maize in eastern Brazil will have moderate changes while Venezuela is expected to have drastic decreases.

Suggested potential adaptation strategies to mitigate the impacts of global warming on agriculture in Latin America include using plant breeding technologies and installing irrigation infrastructure.

Climate Justice and Subsistence Farmers in Latin America

Several studies that investigated the impacts of climate change on agriculture in Latin America suggest that in the poorer countries of Latin America, agriculture composes the most important economic sector and the primary form of sustenance for small farmers. Maize is the only grain still produced as a sustenance crop on small farms in Latin American nations. Scholars argue that the projected decrease of this grain and other crops will threaten the welfare and the economic development of subsistence communities in Latin America. Food security is of particular concern to rural areas that have weak or non-existent food markets to rely on in the case food shortages.

According to scholars who considered the environmental justice implications of climate change, the expected impacts of climate change on subsistence farmers in Latin America and other developing regions are unjust for two reasons. First, subsistence farmers in developing countries, including those in Latin America are disproportionately vulnerable to climate change Second, these nations were the least responsible for causing the problem of anthropogenic induced climate.

According to researchers John F. Morton and T. Roberts, disproportionate vulnerability to climate disasters is socially determined. For example, socioeconomic and policy trends affecting smallholder and subsistence farmers limit their capacity to adapt to change. According to W. Baethgen who studied the vulnerability of Latin American agriculture to climate change, a history of policies and economic dynamics has negatively impacted rural farmers. During the 1950s and through the 1980s, high inflation and appreciated real exchange rates reduced the value of agricultural exports. As a result, farmers in Latin America received lower prices for their products compared to world market prices. Following these outcomes, Latin American policies and national crop programs aimed to stimulate agricultural intensification. These national crop programs benefitted larger commercial farmers more. In the 1980s and 1990s low world market prices for cereals and livestock resulted in decreased agricultural growth and increased rural poverty.

In the book, Fairness in Adaptation to Climate Change, the authors describe the global injustice of climate change between the rich nations of the north, who are the most responsible for global warming and the southern poor countries and minority populations within those countries who are most vulnerable to climate change impacts.

Adaptive planning is challenged by the difficulty of predicting local scale climate change impacts. An expert that considered opportunities for climate change adaptation for rural communities argues that a crucial component to adaptation should include government efforts to lessen the effects of food shortages and famines. This researcher also claims that planning for equitable adaptation and agricultural sustainability will require the engagement of farmers in decision making processes.

North America

A number of studies have been produced which assess the impacts of climate change on agriculture in North America. The IPCC Fourth Assessment Report of agricultural impacts in the region cites 26 different studies. With high confidence, IPCC (2007:14–15) projected that over the first few decades of this century, moderate climate change would increase aggregate yields of rainfed agriculture by 5–20%, but with important variability among regions. Major challenges were projected for crops that are near the warm end of their suitable range or which depend on highly utilized water resources.

Droughts are becoming more frequent and intense in arid and semiarid western North America as temperatures have been rising, advancing the timing and magnitude of spring snow melt floods and reducing river flow volume in summer. Direct effects of climate change include increased heat and water stress, altered crop phenology, and disrupted symbiotic interactions. These effects may be exacerbated by climate changes in river flow, and the combined effects are likely to reduce the abundance of native trees in favor of non-native herbaceous and drought-tolerant competitors, reduce the habitat quality for many native animals, and slow litter decomposition and nutrient cycling. Climate change effects on human water demand and irrigation may intensify these effects.

United States

The US Global Change Research Program (2009) assessed the literature on the impacts of climate change on agriculture in the United States:

- Many crops will benefit from increased atmospheric CO_2 concentrations and low levels of warming, but higher levels of warming will negatively affect growth and yields. Extreme events will likely reduce crop yields.

- Weeds. diseases and insect pests benefit from warming, and will require more attention in regards to pest and weed control.

- Increasing CO_2 concentrations will reduce the land's ability to supply adequate livestock feed. Increased heat, disease, and weather extremes will likely reduce livestock productivity.

According to a paper by Deschenes and Greenstone (2006), predicted increases in temperature and precipitation will have virtually no effect on the most important crops in the US.

Polar Regions (Arctic and Antarctic)

Anisimov *et al..* (2007:655) assessed the literature for the polar region (Arctic and Antarctica). With medium confidence, they concluded that the benefits of a less severe climate were dependent on local conditions. One of these benefits was judged to be increased agricultural and forestry opportunities.

For the *Guardian* newspaper, Brown (2005) reported on how climate change had affected agriculture in Iceland. Rising temperatures had made the widespread sowing of barley possible, which had been untenable twenty years ago. Some of the warming was due to a local (possibly temporary) effect via ocean currents from the Caribbean, which had also affected fish stocks.

Small Islands

In a literature assessment, Mimura *et al.* (2007:689) concluded that on small islands, subsistence and commercial agriculture would very likely be adversely affected by climate change. This projection was assigned "high confidence."

Poverty Impacts

Researchers at the Overseas Development Institute (ODI) have investigated the potential impacts climate change could have on agriculture, and how this would affect attempts at alleviating poverty in the developing world. They argued that the effects from moderate climate change are likely to be mixed for developing countries. However, the vulnerability of the poor in developing countries to short term impacts from climate change, notably the increased frequency and severity of adverse weather events is likely to have a negative impact. This, they say, should be taken into account when defining agricultural policy.

Mitigation and Adaptation in Developing Countries

The Intergovernmental Panel on Climate Change (IPCC) has reported that agriculture is responsible for over a quarter of total global greenhouse gas emissions. Given that agriculture's share in global gross domestic product (GDP) is about 4 percent, these figures suggest that agriculture is highly greenhouse gas intensive. Innovative agricultural practices and technologies can play a role in climate mitigation and adaptation. This adaptation and mitigation potential is nowhere more pronounced than in developing countries where agricultural productivity remains low; poverty, vulnerability and food insecurity remain high; and the direct effects of climate change are expected to be especially harsh. Creating the necessary agricultural technologies and harnessing them to enable developing countries to adapt their agricultural systems to changing climate will require innovations in policy and institutions as well. In this context, institutions and policies are important at multiple scales.

Travis Lybbert and Daniel Sumner suggest six policy principles: (1) The best policy and institutional responses will enhance information flows, incentives and flexibility. (2) Policies and institutions that promote economic development and reduce poverty will often improve agricultural adaptation and may also pave the way for more effective climate change mitigation through agriculture. (3) Business as usual among the world's poor is not adequate. (4) Existing technology options must be made more available and accessible without overlooking complementary capacity and investments. (5) Adaptation and mitigation in agriculture will require local responses, but effective policy responses must also reflect global impacts and inter-linkages. (6) Trade will play a critical role in both mitigation and adaptation, but will itself be shaped importantly by climate change.

The Agricultural Model Intercomparison and Improvement Project (AgMIP) was developed in 2010 to evaluate agricultural models and intercompare their ability to predict climate impacts. In sub-Saharan Africa and South Asia, South America and East Asia, AgMIP regional research teams (RRTs) are conducting integrated assessments to improve understanding of agricultural impacts of climate change (including biophysical and economic impacts) at national and regional scales. Other AgMIP initiatives include global gridded modeling, data and information technology (IT) tool development, simulation of crop pests and diseases, site-based crop-climate sensitivity studies, and aggregation and scaling.

Crop Development Models

Models for climate behavior are frequently inconclusive. In order to further study effects of global warming on agriculture, other types of models, such as *crop development models, yield prediction*, quantities of *water or fertilizer consumed*, can be used. Such models condense the knowledge accumulated of the climate, soil, and effects observed of the results of various agricultural practices. They thus could make it possible to test strategies of adaptation to modifications of the environment.

Because these models are necessarily simplifying natural conditions (often based on the assumption that weeds, disease and insect pests are controlled), it is not clear whether the results they give will have an *in-field* reality. However, some results are partly validated with an increasing number of experimental results.

Other models, such as *insect and disease development* models based on climate projections are also used (for example simulation of aphid reproduction or septoria (cereal fungal disease) development).

Scenarios are used in order to estimate climate changes effects on crop development and yield. Each scenario is defined as a set of meteorological variables, based on generally accepted projections. For example, many models are running simulations based on doubled carbon dioxide projections, temperatures raise ranging from 1 °C up to 5 °C, and with rainfall levels an increase or decrease of 20%. Other parameters may include humidity, wind, and solar activity. Scenarios of crop models are testing farm-level adaptation, such as sowing date shift, climate adapted species (vernalisation need, heat and cold resistance), irrigation and fertilizer adaptation, resistance to disease. Most developed models are about wheat, maize, rice and soybean.

Temperature Potential Effect on Growing Period

Duration of crop growth cycles are above all, related to temperature. An increase in temperature will speed up development. In the case of an annual crop, the duration between sowing and harvesting will shorten (for example, the duration in order to harvest corn could shorten between one and four weeks). The shortening of such a cycle could have an adverse effect on productivity because senescence would occur sooner.

Effect of Elevated Carbon Dioxide on Crops

Carbon dioxide is essential to plant growth. Rising CO_2 concentration in the atmosphere can have both positive and negative consequences.

Increased CO_2 is expected to have positive physiological effects by increasing the rate of photosynthesis. This is known as 'carbon dioxide fertilisation'. Currently, the amount of carbon dioxide in the atmosphere is 380 parts per million. In comparison, the amount of oxygen is 210,000 ppm. This means that often plants may be starved of carbon dioxide as the enzyme that fixes CO_2, RuBisCo, also fixes oxygen in the process of photorespiration. The effects of an increase in carbon dioxide would be higher on C3 crops (such as wheat) than on C4 crops (such as maize), because the former is more susceptible to carbon dioxide shortage. Studies have shown that increased CO_2 leads to fewer stomata developing on plants which leads to reduced water usage. Under optimum conditions of temperature and humidity, the yield increase could reach 36%, if the levels of carbon dioxide are doubled.A study in 2014 posited that CO_2 fertilisation is underestimated due to not explicitly representing CO_2 diffusion inside leaves.

Further, few studies have looked at the impact of elevated carbon dioxide concentrations on whole farming systems. Most models study the relationship between CO_2 and productivity in isolation from other factors associated with climate change, such as an increased frequency of extreme weather events, seasonal shifts, and so on.

In 2005, the Royal Society in London concluded that the purported benefits of elevated carbon dioxide concentrations are "likely to be far lower than previously estimated when factors such as increasing ground-level ozone are taken into account."

Effect on Quality

According to the IPCC's TAR, "The importance of climate change impacts on grain and forage quality emerges from new research. For rice, the amylose content of the grain—a major determinant of cooking quality—is increased under elevated CO_2" (Conroy et al., 1994). Cooked rice grain from plants grown in high-CO_2 environments would be firmer than that from today's plants. However, concentrations of iron and zinc, which are important for human nutrition, would be lower (Seneweera and Conroy, 1997). Moreover, the protein content of the grain decreases under combined increases of temperature and CO_2 (Ziska et al., 1997). Studies using FACE have shown that increases in CO_2 lead to decreased concentrations of micronutrients in crop plants. This may have knock-on effects on other parts of ecosystems as herbivores will need to eat more food to gain the same amount of protein.

Studies have shown that higher CO_2 levels lead to reduced plant uptake of nitrogen (and a smaller number showing the same for trace elements such as zinc) resulting in crops with lower nutritional value. This would primarily impact on populations in poorer countries less able to compensate by eating more food, more varied diets, or possibly taking supplements.

Reduced nitrogen content in grazing plants has also been shown to reduce animal productivity in sheep, which depend on microbes in their gut to digest plants, which in turn depend on nitrogen intake.

Agricultural Surfaces and Climate Changes

Climate change may increase the amount of arable land in high-latitude region by reduction of the amount of frozen lands. A 2005 study reports that temperature in Siberia has increased three degree Celsius in average since 1960 (much more than the rest of the world). However, reports about the impact of global warming on Russian agriculture indicate conflicting probable effects : while they expect a northward extension of farmable lands, they also warn of possible productivity losses and increased risk of drought.

Sea levels are expected to get up to one meter higher by 2100, though this projection is disputed. A rise in the sea level would result in an agricultural land loss, in particular in areas such as South East Asia. Erosion, submergence of shorelines, salinity of the water table due to the increased sea levels, could mainly affect agriculture through inundation of low-lying lands.

Low-lying areas such as Bangladesh, India and Vietnam will experience major loss of rice crop if sea levels rise as expected by the end of the century. Vietnam for example relies heavily on its southern tip, where the Mekong Delta lies, for rice planting. Any rise in sea level of no more than a meter will drown several km^2 of rice paddies, rendering Vietnam incapable of producing its main staple and export of rice.

Erosion and Fertility

The warmer atmospheric temperatures observed over the past decades are expected to lead to a more vigorous hydrological cycle, including more extreme rainfall events. Erosion and soil degradation is more likely to occur. Soil fertility would also be affected by global warming. However, because the ratio of soil organic carbon to nitrogen is mediated by soil biology such that it maintains a narrow range, a doubling of soil organic carbon is likely to imply a doubling in the storage of nitrogen in soils as organic nitrogen, thus providing higher available nutrient levels for plants, supporting higher yield potential. The demand for imported fertilizer nitrogen could decrease, and provide the opportunity for changing costly fertilisation strategies.

Due to the extremes of climate that would result, the increase in precipitations would probably result in greater risks of erosion, whilst at the same time providing soil with better hydration, according to the intensity of the rain. The possible evolution of the organic matter in the soil is a highly contested issue: while the increase in the temperature would induce a greater rate in the production of minerals, lessening the soil organic matter content, the atmospheric CO_2 concentration would tend to increase it.

Potential Effects of Global Climate Change on Pests, Diseases and Weeds

A very important point to consider is that weeds would undergo the same acceleration of cycle as cultivated crops, and would also benefit from carbonaceous fertilization. Since most weeds are C3 plants, they are likely to compete even more than now against C4 crops such as corn. However, on the other hand, some results make it possible to think that weedkillers could increase in effectiveness with the temperature increase.

Global warming would cause an increase in rainfall in some areas, which would lead to an increase of atmospheric humidity and the duration of the wet seasons. Combined with higher temperatures, these could favor the development of fungal diseases. Similarly, because of higher temperatures and humidity, there could be an increased pressure from insects and disease vectors.

Glacier Retreat and Disappearance

The continued retreat of glaciers will have a number of different quantitative impacts. In the areas that are heavily dependent on water runoff from glaciers that melt during the warmer summer months, a continuation of the current retreat will eventually deplete the glacial ice and substantially reduce or eliminate runoff. A reduction in runoff will affect the ability to irrigate crops and will reduce summer stream flows necessary to keep dams and reservoirs replenished.

Approximately 2.4 billion people live in the drainage basin of the Himalayan rivers. India, China, Pakistan, Afghanistan, Bangladesh, Nepal and Myanmar could experience floods followed by severe droughts in coming decades. In India alone, the Ganges provides water for drinking and farming for more than 500 million people. The west coast of North America, which gets much of its water from glaciers in mountain ranges such as the Rocky Mountains and Sierra Nevada, also would be affected.

Ozone and UV-B

Some scientists think agriculture could be affected by any decrease in stratospheric ozone, which could increase biologically dangerous ultraviolet radiation B. Excess ultraviolet radiation B can directly affect plant physiology and cause massive amounts of mutations, and indirectly through changed pollinator behavior, though such changes are not simple to quantify. However, it has not yet been ascertained whether an increase in greenhouse gases would decrease stratospheric ozone levels.

In addition, a possible effect of rising temperatures is significantly higher levels of ground-level ozone, which would substantially lower yields.

ENSO Effects on Agriculture

ENSO (El Niño Southern Oscillation) will affect monsoon patterns more intensely in the future as climate change warms up the ocean's water. Crops that lie on the equatorial belt or under the tropical Walker circulation, such as rice, will be affected by varying monsoon patterns and more unpredictable weather. Scheduled planting and harvesting based on weather patterns will become less effective.

Areas such as Indonesia where the main crop consists of rice will be more vulnerable to the increased intensity of ENSO effects in the future of climate change. University of Washington professor, David Battisti, researched the effects of future ENSO patterns on the Indonesian rice agriculture using [IPCC]'s 2007 annual report and 20 different logistical models mapping out climate factors such as wind pressure, sea-level, and humidity, and found that rice harvest will experience a decrease in yield. Bali and Java, which holds 55% of the rice yields in Indonesia, will be likely to experience 9–10% probably of delayed monsoon patterns, which prolongs the hungry season. Normal planting of rice crops begin in October and harevest by January. However, as climate change affects ENSO and consequently delays planting, harvesting will be late and in drier conditions, resulting in less potential yields.

Impact of Agriculture on Climate Change

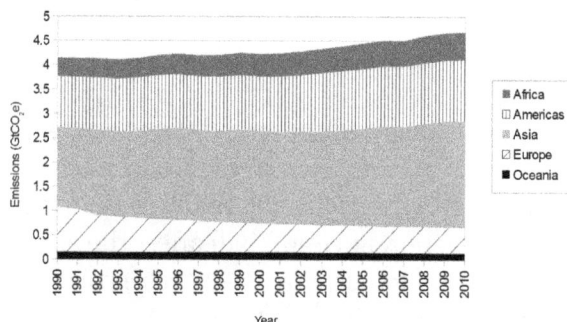

Greenhouse gas emissions from agriculture, by region, 1990-2010.

The agricultural sector is a driving force in the gas emissions and land use effects thought to cause climate change. In addition to being a significant user of land and consumer of fossil fuel, agriculture contributes directly to greenhouse gas emissions through practices such as rice production and the raising of livestock; according to the Intergovernmental Panel on Climate Change, the three main causes of the increase in greenhouse gases observed over the past 250 years have been fossil fuels, land use, and agriculture.

Land Use

Agriculture contributes to greenhouse gas increases through land use in four main ways:

- CO_2 releases linked to deforestation

- Methane releases from rice cultivation

- Methane releases from enteric fermentation in cattle

- Nitrous oxide releases from fertilizer application

Together, these agricultural processes comprise 54% of methane emissions, roughly 80% of nitrous oxide emissions, and virtually all carbon dioxide emissions tied to land use.

The planet's major changes to land cover since 1750 have resulted from deforestation in temperate regions: when forests and woodlands are cleared to make room for fields and pastures, the albedo of the affected area increases, which can result in either warming or cooling effects, depending on local conditions. Deforestation also affects regional carbon reuptake, which can result in increased concentrations of CO_2, the dominant greenhouse gas. Land-clearing methods such as slash and burn compound these effects by burning biomatter, which directly releases greenhouse gases and particulate matter such as soot into the air.

Livestock

Livestock and livestock-related activities such as deforestation and increasingly fuel-intensive farming practices are responsible for over 18% of human-made greenhouse gas emissions, including:

- 9% of global carbon dioxide emissions

- 35–40% of global methane emissions (chiefly due to enteric fermentation and manure)

- 64% of global nitrous oxide emissions (chiefly due to fertilizer use.)

Livestock activities also contribute disproportionately to land-use effects, since crops such as corn and alfalfa are cultivated in order to feed the animals.

In 2010, enteric fermentation accounted for 43% of the total greenhouse gas emissions from all agricultural activity in the world. The meat from ruminants has a higher carbon equivalent footprint than other meats or vegetarian sources of protein based on a global meta-analysis of lifecycle assessment studies. Methane production by animals, principally ruminants, is estimated 15-20% global production of methane.

Worldwide, livestock production occupies 70% of all land used for agriculture, or 30% of the land surface of the Earth.

Environmental Impact of Irrigation

The environmental impacts of irrigation relate to the changes in quantity and quality of soil and water as a result of irrigation and the effects on natural and social conditions in river basins and downstream of an irrigation scheme. The impacts stem from the altered hydrological conditions caused by the installation and operation of the irrigation scheme.

Direct Effects

An irrigation scheme draws water from groundwater, rivers, lakes or overland flow, and distributes it over an area. Hydrological, or direct, effects of doing this include reduction in downstream river flow, increased evaporation in the irrigated area, increased level in the water table as groundwater recharge in the area is increased and flow increased in the irrigated area. Likewise, irrigation has immediate effects on the provision of moisture to the atmosphere, inducing atmospheric instabilities and increasing downwind rainfall, or in other cases modifies the atmospheric circulation, delivering rain to different downwind areas. Increases or decreases in irrigation are a key area of concern in precipitationshed studies, that examine how significant modifications to the delivery of evaporation to the atmosphere can alter downwind rainfall.

Indirect Effects

Indirect effects are those that have consequences that take longer to develop and may also be longer-lasting. The indirect effects of irrigation include the following:

- Waterlogging

- Soil salination

- Ecological damage

- Socioeconomic impacts

The indirect effects of waterlogging and soil salination occur directly on the land being irrigated. The ecological and socioeconomic consequences take longer to happen but can be more far-reaching.

Some irrigation schemes use water wells for irrigation. As a result, the overall water level decreases. This may cause water mining, land/soil subsidence, and, along the coast, saltwater intrusion.

Irrigated land area worldwide occupies about 16% of the total agricultural area and the crop yield of irrigated land is roughly 40% of the total yield. In other words, irrigated land produces 2.5 times more product than non-irrigated land. This article will discuss some of the environmental and socioeconomic impacts of irrigation.

Adverse Impacts

Reduced River Flow

The reduced downstream river flow may cause:

- reduced downstream flooding

- disappearance of ecologically and economically important wetlands or flood forests

- reduced availability of industrial, municipal, household, and drinking water

- reduced shipping routes. Water withdrawal poses a serious threat to the Ganges. In India, barrages control all of the tributaries to the Ganges and divert roughly 60 percent of river flow to irrigation

- reduced fishing opportunities. The Indus River in Pakistan faces scarcity due to over-extraction of water for agriculture. The Indus is inhabited by 25 amphibian species and 147 fish species of which 22 are found nowhere else in the world. It harbors the endangered Indus River dolphin, one of the world's rarest mammals. Fish populations, the main source of protein and overall life support systems for many communities, are also being threatened

- reduced discharge into the sea, which may have various consequences like coastal erosion (e.g. in Ghana) and salt water intrusion in delta's and estuaries. Current water withdrawal from the river Nile for irrigation is so high that, despite its size, in dry periods the river does not reach the sea. The Aral Sea has suffered an "environmental catastrophe" due to the interception of river water for irrigation purposes.

Increased Groundwater Recharge, Waterlogging, Soil Salinity

Increased groundwater recharge stems from the unavoidable deep percolation losses occurring in the irrigation scheme. The lower the irrigation efficiency, the higher the losses. Although fairly high irrigation efficiencies of 70% or more (i.e. losses of 30% or less) can occur with sophisticated techniques like sprinkler irrigation and drip irrigation, or by well managed surface irrigation, in practice the losses are commonly in the order of 40% to 60%. This may cause the following issues:

- rising water tables

- increased storage of groundwater that may be used for irrigation, municipal, household and drinking water by pumping from wells

- waterlogging and drainage problems in villages, agricultural lands, and along roads - with mostly negative consequences. The increased level of the water table can lead to reduced agricultural production.

- shallow water tables - a sign that the aquifer is unable to cope with the groundwater recharge stemming from the deep percolation losses

- where water tables are shallow, the irrigation applications are reduced. As a result, the soil is no longer leached and soil salinity problems develop

- stagnant water tables at the soil surface are known to increase the incidence of water-borne diseases like malaria, filariasis, yellow fever, dengue, and schistosomiasis (Bilharzia) in many areas. Health costs, appraisals of health impacts and mitigation measures are rarely part of irrigation projects, if at all.

- to mitigate the adverse effects of shallow water tables and soil salinization, some form of watertable control, soil salinity control, drainage and drainage system is needed

- as drainage water moves through the soil profile it may dissolve nutrients (either fertilizer-based or naturally occurring) such as nitrates, leading to a buildup of those nutrients in the ground-water aquifer. High nitrate levels in drinking water can be harmful to humans, particularly infants under 6 months, where it is linked to "blue-baby syndrome".

Looking over the shoulder of a Peruvian farmer in the Huarmey delta at waterlogged and salinised irrigated land with a poor crop stand. This illustrates an environmental impact of upstream irrigation developments causing an increased flow of groundwater to this lower-lying area, leading to adverse conditions.

Reduced Downstream River Water Quality

Owing to drainage of surface and groundwater in the project area, which waters may be salinized and polluted by agricultural chemicals like biocides and fertilizers, the quality of the river water below the project area can deteriorate, which makes it less fit for industrial, municipal and household use. It may lead to reduced public health. Polluted river water entering the sea may adversely affect the ecology along the sea shore.

The natural build up of sedimentation can reduce downstream river flows due to the installation of irrigation systems. Sedimentation is an essential part of the ecosystem that requires the natural flux of the river flow. This natural cycle of sediment dispersion replenishes the nutrients in the soil, that will in turn, determine the livelihood of the plants and animals that rely on the sediments carried downstream. The benefits of heavy deposits of sedimentation can be seen in large rivers like the Nile River. The sediment from the delta has built up to form a giant aquifer during flood season, and retains water in the wetlands. The wetlands that are created and sustained due to built up sediment at the basin of the river is a habitat for numerous species of birds However, heavy sedimentation can reduce downstream river water quality and can exacerbate floods up stream. This has been known to happen in the Sanmenxia reservoir in China. The Sanmenxia reservoir is part of a larger man-made project of hydro-electric dams called the Three Gorge Project In 1998, uncertain calculations and heavy sediment greatly affected the reservoir's ability to properly fulfill its flood-control function This also reduces the down stream river water quality. Shifting more towards mass irrigation installments in order to meet more socioeconomic demands is going against the natural balance of nature, and use water pragmatically- use it where it is found

Affected Downstream Water Users

Water becomes scarce for nomadic pastoralist in Baluchistan due to new irrigation developments

Downstream water users often have no legal water rights and may fall victim of the development of irrigation.

Pastoralists and nomadic tribes may find their land and water resources blocked by new irrigation developments without having a legal recourse.

Flood-recession cropping may be seriously affected by the upstream interception of river water for irrigation purposes.

- In Baluchistan, Pakistan, the development of new small-scale irrigation projects depleted the water resources of nomadic tribes traveling annually between Baluchistan and Gujarat or Rajasthan, India

- After the closure of the Kainji dam, Nigeria, 50 to 70 per cent of the downstream area of flood-recession cropping was lost

Lake Manantali, 477 km², displaced 12,000 people.

Lost Land Use Opportunities

Irrigation projects may reduce the fishing opportunities of the original population and the grazing opportunities for cattle. The livestock pressure on the remaining lands may increase considerably, because the ousted traditional pastoralist tribes will have to find their subsistence and existence elsewhere, overgrazing may increase, followed by serious soil erosion and the loss of natural resources. The Manatali reservoir formed by the Manantali dam in Mali intersects the migration routes of nomadic pastoralists and destroyed 43000 ha of savannah, probably leading to overgrazing and erosion elsewhere. Further, the reservoir destroyed 120 km² of forest. The depletion of groundwater aquifers, which is caused by the suppression of the seasonal flood cycle, is damaging the forests downstream of the dam.

Groundwater Mining With Wells, Land Subsidence

Flooding as a consequence of land subsidence

When more groundwater is pumped from wells than replenished, storage of water in the aquifer is being mined and the use of that water is no longer sustainable. As levels fail, it becomes more difficult to extract water and pumps will struggle to maintain the design flow-rate and consume more may fenergy per unit of water. Eventually it may become so difficult to extract groundwater that farmers may be forced to abandon irrigated agriculture. Some notable examples include:

- The hundreds of tubewells installed in the state of Uttar Pradesh, India, with World Bank funding have operating periods of 1.4 to 4.7 hours/day, whereas they were designed to operate 16 hours/day

- In Baluchistan, Pakistan, the development of tubewell irrigation projects was at the expense of the traditional qanat or karez users

- Groundwater-related subsidence of the land due to mining of groundwater occurred in the United States at a rate of 1m for each 13m that the watertable was lowered

- Homes at Greens Bayou near Houston, Texas, where 5 to 7 feet of subsidence has occurred, were flooded during a storm in June 1989 as shown in the picture

Simulation and Prediction

The effects of irrigation on watertable, soil salinity and salinity of drainage and groundwater, and the effects of mitigative measures can be simulated and predicted using agro-hydro-salinity models like SaltMod and SahysMod

Case Studies

1. In India 2.19 million ha have been reported to suffer from waterlogging in irrigation canal commands. Also 3.47 million ha were reported to be seriously salt affected,

2. In the Indus Plains in Pakistan, more than 2 million hectares of land is waterlogged. The soil of 13.6 million hectares within the Gross Command Area was surveyed, which revealed that 3.1 million hectares (23%) was saline. 23% of this was in Sindh and 13% in the Punjab. More than 3 million ha of water-logged lands have been provided with tube-wells and drains at the cost of billions of rupees, but the reclamation objectives were only partially achieved. The Asian Development Bank (ADB) states that 38% of the irrigated area is now waterlogged and 14% of the surface is too saline for use

3. In the Nile delta of Egypt, drainage is being installed in millions of hectares to combat the water-logging resulting from the introduction of massive perennial irrigation after completion of the High Dam at Assuan

4. In Mexico, 15% of the 3 million ha of irrigable land is salinized and 10% is waterlogged

5. In Peru some 0.3 million ha of the 1.05 million ha of irrigable land suffers from degradation.

6. Estimates indicate that roughly one-third of the irrigated land in the major irrigation countries is already badly affected by salinity or is expected to become so in the near future. Present estimates for Israel are 13% of the irrigated land, Australia 20%, China 15%, Iraq 50%, Egypt 30%. Irrigation-induced salinity occurs in large and small irrigation systems alike

7. FAO has estimated that by 1990 about 52 million ha of irrigated land will need to have improved drainage systems installed, much of it subsurface drainage to control salinity

Reduced Downstream Drainage and Groundwater Quality

- The downstream drainage water quality may deteriorate owing to leaching of salts, nutrients, herbicides and pesticides with high salinity and alkalinity. There is threat of soils converting into saline or alkali soils. This may negatively affect the health of the population at the tail-end of the river basin and downstream of the irrigation scheme, as well as the ecological balance. The Aral Sea, for example, is seriously polluted by drainage water.

- The downstream quality of the groundwater may deteriorate in a similar way as the downstream drainage water and have similar consequences

Mitigation of Adverse Effects

Irrigation can have a variety negative impacts on ecology and socioeconomy, which may be mitigated in a number of ways. These include siting the irrigation project in a location which minimises negative impacts. The efficiency of existing projects can be improved and existing degraded croplands can be improved rather than establishing a new irrigation project Developing small-scale, individually owned irrigation systems as an alternative to large-scale, publicly owned and managed schemes. The use of sprinkler irrigation and micro-irrigation systems decrease the risk of waterlogging and erosion. Where practicable, using treated wastewater makes more water available to other users Maintaining flood flows downstream of the dams can ensure that an adequate area is flooded each year, supporting, amongst other objectives, fishery activities.

Delayed Environmental Impacts

It often takes time to accurately predict the impact that new irrigation schemes will have on the ecology and socioeconomy of a region. By the time these predictions are available, a considerable amount of time and resources may have already been expended in the implementation of that project. When that is the case, the project managers will often only change the project if the impact would be considerably more than they had originally expected.

Potential Benefits Outweigh the Potential Disadvantages

Frequently irrigation schemes are seen as extremely necessary for socioeconomic well-being especially in developing countries. One example of this can be demonstrated from a proposal for an irrigation scheme in Malawi. Here it was shown that the potential positive effects of the irrigation project that was being proposed "outweighed the potential negative impacts". It was stated that the impacts would mostly "be localized, minimal, short term occurring during the construction and operation phases of the Project". In order to help alleviate and prevent major environmental impacts, they would use techniques that minimize the potential negative impacts. As far as the region's socioeconomic well-being, there would be no "displacement and/or resettlement envisioned during the implementation of the Project activities". The original primary purposes of the irrigation project were to reduce poverty, improve food security, create local employment, increase household income and enhance the sustainability of land use.

Due to this careful planning this project was successful both in improving the socialeconomic conditions in the region and ensuring that land and water are sustainability into the future.

Environmental Impact of Pesticides

Preparing to spray a hazardous pesticide

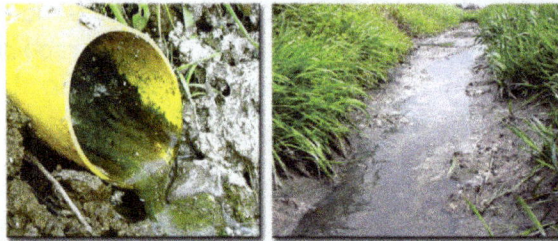

Drainage of fertilizers and pesticides into a stream

The impact of pesticides consists of the effects of pesticides on non-target species. Over 98% of sprayed insecticides and 95% of herbicides reach a destination other than their target species, because they are sprayed or spread across entire agricultural fields. Runoff can carry pesticides into aquatic environments while wind can carry them to other fields, grazing areas, human settlements and undeveloped areas, potentially affecting other species. Other problems emerge from poor production, transport and storage practices. Over time, repeated application increases pest resistance, while its effects on other species can facilitate the pest's resurgence.

Each pesticide or pesticide class comes with a specific set of environmental concerns. Such undesirable effects have led many pesticides to be banned, while regulations have limited and/or reduced the use of others. Over time, pesticides have generally become less persistent and more species-specific, reducing their environmental footprint. In addition the amounts of pesticides applied per hectare have declined, in some cases by 99%. However, the global spread of pesticide use, including the use of older/obsolete pesticides that have been banned in some jurisdictions, has increased overall.

Agriculture and the Environment

The arrival of humans in an area, to live or to conduct agriculture, necessarily has environmental impacts. These range from simple crowding out of wild plants in favor of more desirable cultivars to larger scale impacts such as reducing biodiversity by reducing food availability of native species, which can propagate across food chains. The use of agricultural chemicals such as fertilizer and pesticides magnify those impacts. While advances in agrochemistry have reduced those impacts, for example by the replacement of long-lived chemicals with those that reliably degrade, even in the best case they remain substantial. These effects are magnified by the use of older chemistries and poor management practices.

History

While concern ecotoxicology began with acute poisoning events in the late 19th century; public concern over the undesirable environmental effects of chemicals arose in the early 1960s with the publication of Rachel Carson's book, *Silent Spring*. Shortly thereafter, DDT, originally used to combat malaria, and its metabolites were shown to cause population-level effects in raptorial birds. Initial studies in industrialized countries focused on acute mortality effects mostly involving birds or fish.

Data on pesticide usage remain scattered and/or not publicly available (3). The common practice of incident registration is inadequate for understanding the entirety of effects.

Since 1990, research interest has shifted from documenting incidents and quantifying chemical exposure to studies aimed at linking laboratory, mesocosm and field experiments. The proportion of effect-related publications has increased. Animal studies mostly focus on fish, insects, birds, amphibians and arachnids.

Since 1993, the United States and the European Union have updated pesticide risk assessments, ending the use of acutely toxic organophosphate and carbamate insecticides. Newer pesticides aim at efficiency in target and minimum side effects in nontarget organisms. The phylogenetic proximity of beneficial and pest species complicates the project.

One of the major challenges is to link the results from cellular studies through many levels of increasing complexity to ecosystems.

Specific Pesticide Effects

Pesticide environmental effects	
Pesticide/class	**Effect(s)**
Organochlorine DDT/DDE	Egg shell thinning in raptorial birds
	Endocrine disruptor
	Thyroid disruption properties in rodents, birds, amphibians and fish
	Acute mortality attributed to inhibition of acetylcholine esterase activity
:DDT	Carcinogen

	Endocrine disruptor
:DDT/Diclofol, Dieldrin and Tox-aphene	Juvenile population decline and adult mortality in wildlife reptiles
:DDT/Toxaphene/Parathion	Susceptibility to fungal infection
Triazine	Earthworms became infected with monocystid gregarines
:Chlordane	Interact with vertebrate immune systems
Carbamates, the phenoxy herbicide 2,4-D, and atrazine	Interact with vertebrate immune systems
Anticholinesterase	Bird poisoning
	Animal infections, disease outbreaks and higher mortality.
Organophosphate	Thyroid disruption properties in rodents, birds, amphibians and fish
	Acute mortality attributed to inhibition of acetylcholine esterase activity
	Immunotoxicity, primarily caused by the inhibition of serine hydrolases or esterases
	Oxidative damage
	Modulation of signal transduction pathways
	Impaired metabolic functions such as thermoregulation, water and/or food intake and behavior, impaired development, reduced reproduction and hatching success in vertebrates.
Carbamate	Thyroid disruption properties in rodents, birds, amphibians and fish
	Impaired metabolic functions such as thermoregulation, water and/or food intake and behavior, impaired development, reduced reproduction and hatching success in vertebrates.
	Interact with vertebrate immune systems
	Acute mortality attributed to inhibition of acetylcholine esterase activity
Phenoxy herbicide 2,4-D	Interact with vertebrate immune systems
Atrazine	Interact with vertebrate immune systems
	Reduced northern leopard frog (Rana pipiens) populations because atrazine killed phytoplankton, thus allowing light to penetrate the water column and periphyton to assimilate nutrients released from the plankton. Periphyton growth provided more food to grazers, increasing snail populations, which provide intermediate hosts for trematode.

Pyrethroid	Thyroid disruption properties in rodents, birds, amphibians and fish
Thiocarbamate	Thyroid disruption properties in rodents, birds, amphibians and fish
Triazine	Thyroid disruption properties in rodents, birds, amphibians and fish
Triazole	Thyroid disruption properties in rodents, birds, amphibians and fish
	Impaired metabolic functions such as thermoregulation, water and/or food intake and behavior, impaired development, reduced reproduction and hatching success in vertebrates.
Nicotinoid	respiratory, cardiovascular, neurological, and immunological toxicity in rats and humans
	Disrupt biogenic amine signaling and cause subsequent olfactory dysfunction, as well as affecting foraging behavior, learning and memory.
Imidacloprid, Imidacloprid/pyrethroid λ-cyhalothrin	Impaired foraging, brood development, and colony success in terms of growth rate and new queen production.
Thiamethoxa	High honey bee worker mortality due to homing failure (risks for colony collapse remain controversial)
Spinosyns	Affect various physiological and behavioral traits of beneficial arthropods, particularly hymenopterans
Bt corn/Cry	Reduced abundance of some insect taxa, predominantly susceptible Lepidopteran herbivores as well as their predators and parasitoids.
Herbicide	Reduced food availability and adverse secondary effects on soil invertebrates and butterflies
	Decreased species abundance and diversity in small mammals.
Benomyl	Altered the patch-level floral display and later a two-thirds reduction of the total number of bee visits and in a shift in the visitors from large-bodied bees to small-bodied bees and flies
Herbicide and planting cycles	Reduced survival and reproductive rates in seed-eating or carnivorous birds

Air

Spraying a mosquito pesticide over a city

Pesticides can contribute to air pollution. Pesticide drift occurs when pesticides suspended in the air as particles are carried by wind to other areas, potentially contaminating them. Pesticides that are applied to crops can volatilize and may be blown by winds into nearby areas, potentially posing a threat to wildlife. Weather conditions at the time of application as well as temperature and relative humidity change the spread of the pesticide in the air. As wind velocity increases so does the spray drift and exposure. Low relative humidity and high temperature result in more spray evaporating. The amount of inhalable pesticides in the outdoor environment is therefore often dependent on the season. Also, droplets of sprayed pesticides or particles from pesticides applied as dusts may travel on the wind to other areas, or pesticides may adhere to particles that blow in the wind, such as dust particles. Ground spraying produces less pesticide drift than aerial spraying does. Farmers can employ a buffer zone around their crop, consisting of empty land or non-crop plants such as evergreen trees to serve as windbreaks and absorb the pesticides, preventing drift into other areas. Such windbreaks are legally required in the Netherlands.

Pesticides that are sprayed on to fields and used to fumigate soil can give off chemicals called volatile organic compounds, which can react with other chemicals and form a pollutant called tropospheric ozone. Pesticide use accounts for about 6 percent of total tropospheric ozone levels.

Water

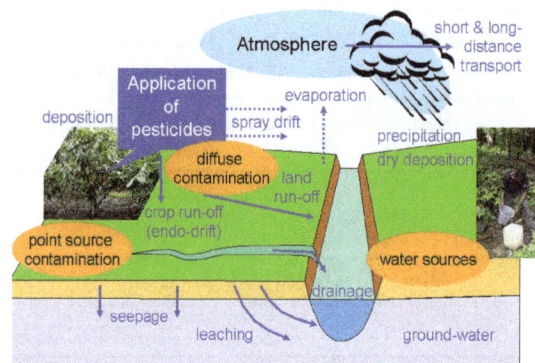

Pesticide pathways

In the United States, pesticides were found to pollute every stream and over 90% of wells sampled in a study by the US Geological Survey. Pesticide residues have also been found in rain and groundwater. Studies by the UK government showed that pesticide concentrations exceeded those allowable for drinking water in some samples of river water and groundwater.

Pesticide impacts on aquatic systems are often studied using a hydrology transport model to study movement and fate of chemicals in rivers and streams. As early as the 1970s quantitative analysis of pesticide runoff was conducted in order to predict amounts of pesticide that would reach surface waters.

There are four major routes through which pesticides reach the water: it may drift outside of the intended area when it is sprayed, it may percolate, or leach, through the soil, it may be carried to the water as runoff, or it may be spilled, for example accidentally or through neglect. They may also be carried to water by eroding soil. Factors that affect a pesticide's ability to contaminate water include its water solubility, the distance from an application site to a body of water, weather, soil type, presence of a growing crop, and the method used to apply the chemical.

Maximum limits of allowable concentrations for individual pesticides in public bodies of water are set by the Environmental Protection Agency in the US. Similarly, the government of the United Kingdom sets Environmental Quality Standards (EQS), or maximum allowable concentrations of some pesticides in bodies of water above which toxicity may occur. The European Union also regulates maximum concentrations of pesticides in water.

Soil

Caution against entering a field sprayed with sulphuric acid

Many of the chemicals used in pesticides are persistent soil contaminants, whose impact may endure for decades and adversely affect soil conservation.

The use of pesticides decreases the general biodiversity in the soil. Not using the chemicals results in higher soil quality, with the additional effect that more organic matter in the soil allows for higher water retention. This helps increase yields for farms in drought years, when organic farms have had yields 20-40% higher than their conventional counterparts. A smaller content of organic matter in the soil increases the amount of pesticide that will leave the area of application, because organic matter binds to and helps break down pesticides.

Degradation and sorption are both factors which influence the persistence of pesticides in soil. Depending on the chemical nature of the pesticide, such processes control directly the transportation from soil to water, and in turn to air and our food. Breaking down organic substances, degradation, involves interactions among microorganisms in the soil. Sorption affects bioaccumulation of pesticides which are dependent on organic matter in the soil. Weak organic acids have been shown to be weakly sorbed by soil, because of pH and mostly acidic structure. Sorbed chemicals have been shown to be less accessible to microorganisms. Aging mechanisms are poorly understood but as residence times in soil increase, pesticide residues become more resistant to degradation and extraction as they lose biological activity.

Effect on Plants

Nitrogen fixation, which is required for the growth of higher plants, is hindered by pesticides in soil. The insecticides DDT, methyl parathion, and especially pentachlorophenol have been shown to interfere with legume-rhizobium chemical signaling. Reduction of this symbiotic chemical signaling results in reduced nitrogen fixation and thus reduced crop yields. Root nodule formation in these plants saves the world economy $10 billion in synthetic nitrogen fertilizer every year.

Crop spraying

Pesticides can kill bees and are strongly implicated in pollinator decline, the loss of species that pollinate plants, including through the mechanism of Colony Collapse Disorder,in which worker bees from a beehive or western honey bee colony abruptly disappear. Application of pesticides to crops that are in bloom can kill honeybees, which act as pollinators. The USDA and USFWS estimate that US farmers lose at least $200 million a year from reduced crop pollination because pesticides applied to fields eliminate about a fifth of honeybee colonies in the US and harm an additional 15%.

On the other side, pesticides have some direct harmful effect on plant including poor root hair development, shoot yellowing and reduced plant growth.

Effect on Animals

Many kinds of animals are harmed by pesticides, leading many countries to regulate pesticide usage through Biodiversity Action Plans.

Animals including humans may be poisoned by pesticide residues that remain on food, for example when wild animals enter sprayed fields or nearby areas shortly after spraying.

Pesticides can eliminate some animals' essential food sources, causing the animals to relocate, change their diet or starve. Residues can travel up the food chain; for example, birds can be harmed when they eat insects and worms that have consumed pesticides. Earthworms digest organic matter and increase nutrient content in the top layer of soil. They protect human health by ingesting decomposing litter and serving as bioindicators of soil activity. Pesticides have had harmful effects on growth and reproduction on earthworms. Some pesticides can bioaccumulate, or build up to toxic levels in the bodies of organisms that consume them over time, a phenomenon that impacts species high on the food chain especially hard.

Birds

In England, the use of pesticides in gardens and farmland has seen a reduction in the number of common chaffinches

The US Fish and Wildlife Service estimates that 72 million birds are killed by pesticides in the United States each year. Bald eagles are common examples of nontarget organisms that are impacted by pesticide use. Rachel Carson's book *Silent Spring* dealt with damage to bird species due to pesticide bioaccumulation. There is evidence that birds are continuing to be harmed by pesticide use. In the farmland of the United Kingdom, populations of ten different bird species declined by 10 million breeding individuals between 1979 and 1999, allegedly from loss of plant and invertebrate species on which the birds feed. Throughout Europe, 116 species of birds were threatened as of 1999. Reductions in bird populations have been found to be associated with times and areas in which pesticides are used. DDE-induced egg shell thinning has especially affected European and North American bird populations. In another example, some types of fungicides used in peanut farming are only slightly toxic to birds and mammals, but may kill earthworms, which can in turn reduce populations of the birds and mammals that feed on them.

Some pesticides come in granular form. Wildlife may eat the granules, mistaking them for grains of food. A few granules of a pesticide may be enough to kill a small bird.

The herbicide paraquat, when sprayed onto bird eggs, causes growth abnormalities in embryos and reduces the number of chicks that hatch successfully, but most herbicides do not directly cause much harm to birds. Herbicides may endanger bird populations by reducing their habitat.

Aquatic Life

Using an aquatic herbicide

Wide field margins can reduce fertilizer and pesticide pollution in streams and rivers

Fish and other aquatic biota may be harmed by pesticide-contaminated water. Pesticide surface runoff into rivers and streams can be highly lethal to aquatic life, sometimes killing all the fish in a particular stream.

Application of herbicides to bodies of water can cause fish kills when the dead plants decay and consume the water's oxygen, suffocating the fish. Herbicides such as copper sulfite that are applied to water to kill plants are toxic to fish and other water animals at concentrations similar to those used to kill the plants. Repeated exposure to sublethal doses of some pesticides can cause physiological and behavioral changes that reduce fish populations, such as abandonment of nests and broods, decreased immunity to disease and decreased predator avoidance.

Application of herbicides to bodies of water can kill plants on which fish depend for their habitat.

Pesticides can accumulate in bodies of water to levels that kill off zooplankton, the main source of food for young fish. Pesticides can also kill off insects on which some fish feed, causing the fish to travel farther in search of food and exposing them to greater risk from predators.

The faster a given pesticide breaks down in the environment, the less threat it poses to aquatic life. Insecticides are typically more toxic to aquatic life than herbicides and fungicides.

Amphibians

In the past several decades, amphibian populations have declined across the world, for unexplained reasons which are thought to be varied but of which pesticides may be a part.

Pesticide mixtures appear to have a cumulative toxic effect on frogs. Tadpoles from ponds containing multiple pesticides take longer to metamorphose and are smaller when they do, decreasing their ability to catch prey and avoid predators. Exposing tadpoles to the organochloride endosulfan at levels likely to be found in habitats near fields sprayed with the chemical kills the tadpoles and causes behavioral and growth abnormalities.

The herbicide atrazine can turn male frogs into hermaphrodites, decreasing their ability to reproduce. Both reproductive and nonreproductive effects in aquatic reptiles and amphibians have been reported. Crocodiles, many turtle species and some lizards lack sex-distinct chromosomes until after fertilization during organogenesis, depending on temperature. Embryonic exposure in turtles to various PCBs causes a sex reversal. Across the United States and Canada disorders such as decreased hatching success, feminization, skin lesions, and other developmental abnormalities have been reported.

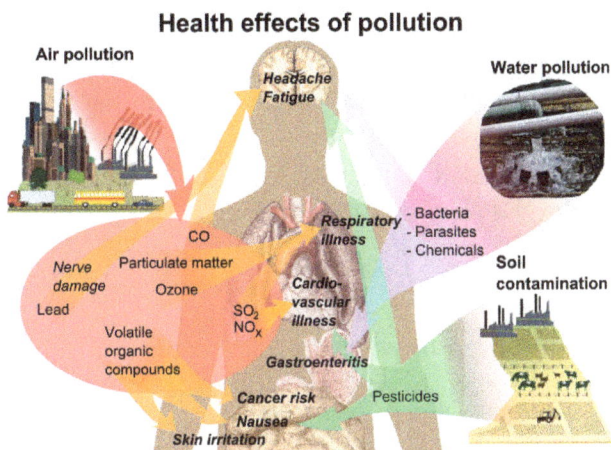

Pesticides are implicated in a range of impacts on human health due to pollution

Humans

Pesticides can enter the body through inhalation of aerosols, dust and vapor that contain pesticides; through oral exposure by consuming food/water; and through skin exposure by direct contact. Pesticides secrete into soils and groundwater which can end up in drinking water, and pesticide spray can drift and pollute the air.

The effects of pesticides on human health depend on the toxicity of the chemical and the length and magnitude of exposure. Farm workers and their families experience the greatest exposure to agricultural pesticides through direct contact. Every human contains pesticides in their fat cells.

Children are more susceptible and sensitive to pesticides, because they are still developing and have a weaker immune system than adults. Children may be more exposed due to their closer proximity to the ground and tendency to put unfamiliar objects in their mouth. Hand to mouth contact depends on the child's age, much like lead exposure. Children under the age of six months are more apt to experience exposure from breast milk and inhalation of small particles. Pesticides tracked into the home from family members increase the risk of exposure. Toxic residue in food may contribute to a child's exposure. The chemicals can bioaccumulate in the body over time.

Exposure effects can range from mild skin irritation to birth defects, tumors, genetic changes, blood and nerve disorders, endocrine disruption, coma or death. Developmental effects have been associated with pesticides. Recent increases in childhood cancers in throughout North America, such as leukemia, may be a result of somatic cell mutations. Insecticides targeted to disrupt insects can have harmful effects on mammalian nervous systems. Both chronic and acute alterations have been observed in exposees. DDT and its breakdown product DDE disturb estrogenic activity and possibly lead to breast cancer. Fetal DDT exposure reduces male penis size in animals and can produce undescended testicles. Pesticide can affect fetuses in early stages of development, in utero and even if a parent was exposed before conception. Reproductive disruption has the potential to occur by chemical reactivity and through structural changes.

Persistent Organic Pollutants

Persistent organic pollutants (POPs) are compounds that resist degradation and thus remain in the environment for years. Some pesticides, including aldrin, chlordane, DDT, dieldrin, endrin, heptachlor, hexachlorobenzene, mirex and toxaphene, are considered POPs. Some POPs have the ability to volatilize and travel great distances through the atmosphere to become deposited in remote regions. Such chemicals may have the ability to bioaccumulate and biomagnify and can bioconcentrate (i.e. become more concentrated) up to 70,000 times their original concentrations. POPs can affect non-target organisms in the environment and increase risk to humans by disruption in the endocrine, reproductive, and immune systems.

Pest Resistance

Pests may evolve to become resistant to pesticides. Many pests will initially be very susceptible to pesticides, but following mutations in their genetic makeup become resistant and survive to reproduce.

Resistance is commonly managed through pesticide rotation, which involves alternating among pesticide classes with different modes of action to delay the onset of or mitigate existing pest resistance.

Pest Rebound and Secondary Pest Outbreaks

Non-target organisms can also be impacted by pesticides. In some cases, a pest insect that is controlled by a beneficial predator or parasite can flourish should an insecticide application kill both pest and beneficial populations. A study comparing biological pest control and pyrethroid insecticide for diamondback moths, a major cabbage family insect pest, showed that the pest population rebounded due to loss of insect predators, whereas the biocontrol did not show the same effect. Likewise, pesticides sprayed to control mosquitoes may temporarily depress mosquito populations, however they may result in a larger population in the long run by damaging natural controls. This phenomenon, wherein the population of a pest species rebounds to equal or greater numbers than it had before pesticide use, is called pest resurgence and can be linked to elimination of its predators and other natural enemies.

Loss of predator species can also lead to a related phenomenon called secondary pest outbreaks, an increase in problems from species that were not originally a problem due to loss of their predators or parasites. An estimated third of the 300 most damaging insects in the US were originally secondary pests and only became a major problem after the use of pesticides. In both pest resurgence and secondary outbreaks, their natural enemies were more susceptible to the pesticides than the pests themselves, in some cases causing the pest population to be higher than it was before the use of pesticide.

Eliminating Pesticides

Many alternatives are available to reduce the effects pesticides have on the environment. Alternatives include manual removal, applying heat, covering weeds with plastic, placing traps and lures, removing pest breeding sites, maintaining healthy soils that breed healthy, more resistant plants, cropping native species that are naturally more resistant to native pests and supporting biocontrol agents such as birds and other pest predators. In the United States, conventional pesticide use peaked in 1979, and by 2007, had been reduced by 25 percent from the 1979 peak level, while US agricultural output increased by 43 percent over the same period.

Biological controls such as resistant plant varieties and the use of pheromones, have been successful and at times permanently resolve a pest problem. Integrated Pest Management (IPM) employs chemical use only when other alternatives are ineffective. IPM causes less harm to humans and the environment. The focus is broader than on a specific pest, considering a range of pest control alternatives. Biotechnology can also be an innovative way to control pests. Strains can be genetically modified (GM) to increase their resistance to pests. However the same techniques can be used to increase pesticide resistance and was employed by Monsanto to create glyphosate-resistant strains of major crops. In 2010, 70% of all the corn that was planted was resistant to glyphosate; 78% of cotton, and 93% of all soybeans.

Plasticulture

Plastic mulch used for growing strawberries.

The term plasticulture refers to the practice of using plastic materials in agricultural applications.

The plastic materials themselves are often and broadly referred to as "ag plastics." Plasticulture ag plastics include soil fumigation film, irrigation drip tape/tubing, nursery pots and silage bags, but the term is most often used to describe all kinds of plastic plant/soil coverings. Such coverings range from plastic mulch film, row coverings, high and low tunnels (polytunnels), to plastic greenhouses.

Polyethylene (PE) is the plastic film used by the majority of growers because of its affordability, flexibility and easy manufacturing. It comes in a variety of thicknesses, such as a low density form (LDPE) as well as a linear low density form (LLDPE). These can be modified by addition of certain elements to the plastic that give it properties beneficial to plant growth such as reduced water loss, UV stabilization to cool soil and prevent insects, elimination of Photosynthetically active radiation to prevent weed growth, IR opacity, antidrip/antifog, and fluorescent films.

Greenhouses and Walk-in Tunnel Covers

A Greenhouse is a large structure in which it is possible to stand and work with automated ventilation. High tunnels are hoop houses, manually ventilated by rolling up the sides. Greenhouse and high tunnel films are usually within the parameters of 80-220μm thick and 20m wide, and have a life span between 6–45 months dependant on several factors. Monolayer polyethylene films are better suited for less extreme environmental conditions, while multilayer covers made of three layers, one EVA19 layer inserted between two low-density polyethylene layers has been shown to have a better performance under harsh conditions.

Small Tunnel Covers

Small tunnel covers are about 1m wide and 1m high, and have a thinner polyethylene film than the large tunnel covers, usually below 80μm. Their lifetime is also shorter than that of the larger versions, they usually have a duration of 6–8 months. Use of small tunnels is less popular than both the more expensive but durable greenhouses/walk-in tunnels and the cheaper plastic mulch.

Plastic Mulch

Mulching is when a thin plastic film is placed over the ground, poking holes at regular intervals for seeds to be planted in, or placing it directly over plants in the beginning stages of growth. The films remain in place for the duration of the cultivation (usually 2–4 months) and usually have a thickness of 12-80μm. The main functions of plastic mulch are to insulate and maintain a consistent temperature and humidity of the soil, preventing evaporation of moisture from the soil, minimization of seedtime and harvest, prevent weed growth, and to prevent erosion. Pigmented or colourless films can be used, each with specific advantages and disadvantages over the other.

Black films prevent weed growth, but do not transmit light to heat up the soil; clear films transmit light and heat the soil, but promote weed growth. Photosensitive films have been developed that are pigmented to prevent weed growth, but still transmit light to heat the soil. These photosensitive films are more costly than either the clear or black polyethylene sheeting. In a study using okra, the total yield of the okra and fruit number was two times more when using plastic mulching than without mulching. They found that black plastic mulch controls evaporation from the soil, improves soil water retention. Plastic mulching proved to reduce irrigation requirements in pepper by 14-29% because of elimination of soil evaporation.

Flowering time was also reduced in okra when black plastic mulch was used; the plants reached 50% flowering 3–6 days earlier than un-mulched plots. Plant height in okra was significantly increased with black plastic mulch use compared to those grown in bare soil. Evaporation from soil accounts for 25-50% of water used in irrigation, using plastic mulch prevents much of this evaporation and thus reduces the amount of water needed to grow the crop. This conservation of water makes plastic mulch favourable for farmers in dry and arid climates where water is a limited resource. As the second most used ag plastic in the world, the volume of plastic mulch used every year is estimated at 700,000t.

Origins and Development Around the World

The first use of plastic film in agriculture was to in an effort to make a cheaper version of a glasshouse. In 1948 Professor E.M. Emmert built the first plastic greenhouse, a wooden structure covered with cellulose acetate film. He later switched this to a more effective polyethylene film. After this introduction of plastic film to agriculture it began being used at a larger scale around the world by the early 1950s to replace paper for mulching vegetables.

By 1999 almost 30 million acres worldwide were covered in plastic mulch. Only a small percentage of this was in the United States (185 000 acres), the majority of this plastic growth was happening in economically poor areas of the world and previously unproductive desert regions, such as Almeria in southern Spain.

The largest concentrations of greenhouses around the world are mainly found in two areas, with 80% throughout the Far East (China, Japan, Korea), and 15% in the Mediterranean basin. The area of greenhouse cover is still increasing at a fast rate, during the last decade it is estimated that it has been growing by 20% every year. Areas such as the Middle East and Africa are growing in their use of plastic greenhouses by 15-20% per year, compared to the weak growth in more developed and economically stable areas such as Europe. China leads the world's growth at 30% per year, translating into a volume of plastic film reaching 1,000,000 t/year. In 2006 80% of the area covered by plastic mulch is found in China where it has a growth rate of 25% per year; this is the highest in the world.

Since its introduction in the 1950s, plastic film has been designed and developed to increase produce yield, increase produce size and shorten growth time. Developments in plastic film include durability, optical (ultraviolet, visible, near infrared, and middle infrared) properties, and the antidrip or antifog effect. Recent developments in this area include UV-blocking, NIR-blocking, fluorescent, and ultrathermic films.

Large-scale Usage in Southern Spain

The "sea of plastic" covering 20,000 ha of the Campo de Dalías around El Ejido and Roquetas de Mar in southern Spain.

The use of plasticulture in agriculture is growing rapidly, perhaps nowhere more visibly than around Almería in southern Spain. The eastern approaches to Almería, north of the airport, are densely covered, as is a large area further north-east, surrounding the towns of Campohermoso, Los Pipaces and Los Grillos (close to Níjar).

The densest concentration lies about 20 km south-west of Almería, where almost the entire Campo de Dalías, a low-lying cape, is now under plastic (an estimated area of 20,000 hectares). Further west, a similar, but smaller, coastal plain around Carchuna, southeast of Motril, is similarly enveloped. The technique is not restricted to the plains; it is also applied to wide terraces on the sides of shallow valleys, as the valley north of Castell de Ferro shows.

Elsewhere along the Costa Tropical and the Costa del Sol, particularly between Almería and Málaga, fruit trees growing on terraces in steeper valleys may be covered with vast tents of plastic netting.

Aerial photographs and Coordinates

- Centre of Campo de Dalías - Google maps Dense concentration of plastic greenhouses

- 36°52′N 2°23′W36.86°N 2.38°W Eastern outskirts of Almería

- 36°55′N 2°09′W36.91°N 2.15°W Níjar valley

- 36°42′N 3°26′W36.70°N 3.44°W Carchuna

- 36°44′24″N 3°21′54″W36.74°N 3.365°W Valley north of Castell de Ferro

Environmental Aspects

Recycling

One significant component of plasticulture is the disposal of used ag plastics. Technologies exist which allow for many ag plastics to be recycled into viable plastic resins for reuse in the plastics manufacturing industry.

References

- George Tyler Miller (1 January 2004). Sustaining the Earth: An Integrated Approach. Thomson/Brooks/Cole. pp. 211–216. ISBN 978-0-534-40088-0.

- Howell V. Daly; John T. Doyen; Alexander H. Purcell (1 January 1998). Introduction to Insect Biology and Diversity. Oxford University Press. pp. 279–300. ISBN 978-0-19-510033-4.

- N.T. Singh, 2005. Irrigation and soil salinity in the Indian subcontinent: past and present. Lehigh University Press. ISBN 0-934223-78-5, ISBN 978-0-934223-78-2, 404 p.

- Hemphill, Delbert (March 1993). "Agricultural Plastics as Solid Waste: What are the Options for Disposal?" (PDF). Hort Technology. 3 (1): 70–73. Retrieved 23 April 2015.

- Kidd, Greg (1999–2000). "Pesticides and Plastic Mulch Threaten the Health of Maryland and Virginia East Shore Waters" (PDF). Pesticides and You. 19 (4): 22–23. Retrieved 23 April 2015.

- Cline, W.R. (March 2008), "Global Warming and Agriculture", Finance and Development, International Monetary Fund, 45 (1), Archived from the original on 17 August 2014 . Archived

- Lobell, D.; et al. (2008b), Prioritizing climate change adaptation needs for food security - Policy Brief, Center on Food Security and the Environment, Stanford University. Archived 27 September 2014.

- Non-Annex I NC (11 December 2014), Non-Annex I national communications, United Nations Framework Convention on Climate Change, Archived from the original on 13 September 2014 . Archived .

- Lorenz, Eric S. (2009). "Potential Health Effects of Pesticides" (PDF). Ag Communications and Marketing: 1–8. Retrieved February 2014. Check date values in: |access-date= (help)

- "Irrigation potential in Africa: A basin approach". Natural Resources Management and Environment Department. Retrieved 13 March 2014.

- Lamberth, C.; Jeanmart, S.; Luksch, T.; Plant, A. (2013). "Current Challenges and Trends in the Discovery of Agrochemicals". Science. 341 (6147): 742–6. doi:10.1126/science.1237227. PMID 23950530.

- Kohler, H. -R.; Triebskorn, R. (2013). "Wildlife Ecotoxicology of Pesticides: Can We Track Effects to the Population Level and Beyond?". Science. 341 (6147): 759–765. doi:10.1126/science.1237591.

7

Technological Aspects of Crop Production

Technological innovation has played a very important role in the productivity of crops. Seeds and crops can be genetically altered to increase yield, growth rate as well as withstand disease and pests. This chapter elucidates the changes that have been made possible by technology in agriculture.

Genetically Modified Crops

Genetically modified crops (GMCs, GM crops, or biotech crops) are plants used in agriculture, the DNA of which has been modified using genetic engineering techniques. In most cases, the aim is to introduce a new trait to the plant which does not occur naturally in the species. Examples in food crops include resistance to certain pests, diseases, or environmental conditions, reduction of spoilage, or resistance to chemical treatments (e.g. resistance to a herbicide), or improving the nutrient profile of the crop. Examples in non-food crops include production of pharmaceutical agents, biofuels, and other industrially useful goods, as well as for bioremediation.

Farmers have widely adopted GM technology. Between 1996 and 2015, the total surface area of land cultivated with GM crops increased by a factor of 100, from 17,000 km^2 (4.2 million acres) to 1,797,000 km^2 (444 million acres). 10% of the world's arable land was planted with GM crops in 2010. In the US, by 2014, 94% of the planted area of soybeans, 96% of cotton and 93% of corn were genetically modified varieties. Use of GM crops expanded rapidly in developing countries, with about 18 million farmers growing 54% of worldwide GM crops by 2013. A 2014 meta-analysis concluded that GM technology adoption had reduced chemical pesticide use by 37%, increased crop yields by 22%, and increased farmer profits by 68%. This reduction in pesticide use has been ecologically beneficial, but benefits may be reduced by overuse. Yield gains and pesticide reductions are larger for insect-resistant crops than for herbicide-tolerant crops. Yield and profit gains are higher in developing countries than in developed countries.

There is a scientific consensus that currently available food derived from GM crops poses no greater risk to human health than conventional food, but that each GM food needs to be tested on a case-by-case basis before introduction. Nonetheless, members of the public are much less likely than scientists to perceive GM foods as safe. The legal and regulatory status of GM foods varies by country, with some nations banning or restricting them, and others permitting them with widely differing degrees of regulation.

However, opponents have objected to GM crops on several grounds, including environmental concerns, whether food produced from GM crops is safe, whether GM crops are needed to address the world's food needs, and concerns raised by the fact these organisms are subject to intellectual property law.

Gene Transfer in Nature and Traditional Agriculture

DNA transfers naturally between organisms. Several natural mechanisms allow gene flow across species. These occur in nature on a large scale – for example, it is one mechanism for the development of antibiotic resistance in bacteria. This is facilitated by transposons, retrotransposons, proviruses and other mobile genetic elements that naturally translocate DNA to new loci in a genome. Movement occurs over an evolutionary time scale.

The introduction of foreign germplasm into crops has been achieved by traditional crop breeders by overcoming species barriers. A hybrid cereal grain was created in 1875, by crossing wheat and rye. Since then important traits including dwarfing genes and rust resistance have been introduced. Plant tissue culture and deliberate mutations have enabled humans to alter the makeup of plant genomes.

History

The first genetically modified crop plant was produced in 1982, an antibiotic-resistant tobacco plant. The first field trials occurred in France and the USA in 1986, when tobacco plants were engineered for herbicide resistance. In 1987, Plant Genetic Systems (Ghent, Belgium), founded by Marc Van Montagu and Jeff Schell, was the first company to genetically engineer insect-resistant (tobacco) plants by incorporating genes that produced insecticidal proteins from *Bacillus thuringiensis* (Bt).

The People's Republic of China was the first country to allow commercialized transgenic plants, introducing a virus-resistant tobacco in 1992, which was withdrawn in 1997. The first genetically modified crop approved for sale in the U.S., in 1994, was the *FlavrSavr* tomato. It had a longer shelf life, because it took longer to soften after ripening. In 1994, the European Union approved tobacco engineered to be resistant to the herbicide bromoxynil, making it the first commercially genetically engineered crop marketed in Europe.

In 1995, Bt Potato was approved by the US Environmental Protection Agency, making it the country's first pesticide producing crop. In 1995 canola with modified oil composition (Calgene), Bt maize (Ciba-Geigy), bromoxynil-resistant cotton (Calgene), Bt cotton (Monsanto), glyphosate-resistant soybeans (Monsanto), virus-resistant squash (Asgrow), and additional delayed ripening tomatoes (DNAP, Zeneca/Peto, and Monsanto) were approved. As of mid-1996, a total of 35 approvals had been granted to commercially grow 8 transgenic crops and one flower crop (carnation), with 8 different traits in 6 countries plus the EU. In 2000, Vitamin A-enriched golden rice was developed, though as of 2016 it was not yet in commercial production. In 2013 the leaders of the three research teams that first applied genetic engineering to crops, Robert Fraley, Marc Van Montagu and Mary-Dell Chilton were awarded the World Food Prize for improving the "quality, quantity or availability" of food in the world.

Methods

Genetically engineered crops have genes added or removed using genetic engineering techniques, originally including gene guns, electroporation, microinjection and agrobacterium. More recently, CRISPR and TALEN offered much more precise and convenient editing techniques.

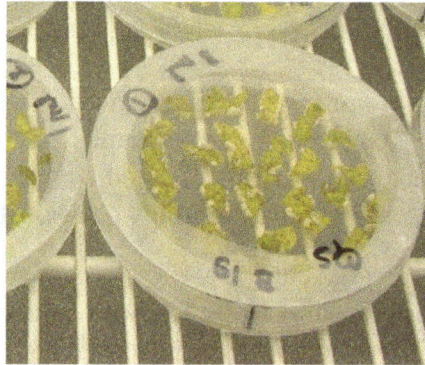

Plants (Solanum chacoense) being transformed using agrobacterium

Gene guns (also known as biolistics) "shoot" (direct high energy particles or radiations against) target genes into plant cells. It is the most common method. DNA is bound to tiny particles of gold or tungsten which are subsequently shot into plant tissue or single plant cells under high pressure. The accelerated particles penetrate both the cell wall and membranes. The DNA separates from the metal and is integrated into plant DNA inside the nucleus. This method has been applied successfully for many cultivated crops, especially monocots like wheat or maize, for which transformation using *Agrobacterium tumefaciens* has been less successful. The major disadvantage of this procedure is that serious damage can be done to the cellular tissue.

Agrobacterium tumefaciens-mediated transformation is another common technique. Agrobacteria are natural plant parasites, and their natural ability to transfer genes provides another engineering method. To create a suitable environment for themselves, these Agrobacteria insert their genes into plant hosts, resulting in a proliferation of modified plant cells near the soil level (crown gall). The genetic information for tumour growth is encoded on a mobile, circular DNA fragment (plasmid). When *Agrobacterium* infects a plant, it transfers this T-DNA to a random site in the plant genome. When used in genetic engineering the bacterial T-DNA is removed from the bacterial plasmid and replaced with the desired foreign gene. The bacterium is a vector, enabling transportation of foreign genes into plants. This method works especially well for dicotyledonous plants like potatoes, tomatoes, and tobacco. Agrobacteria infection is less successful in crops like wheat and maize.

Electroporation is used when the plant tissue does not contain cell walls. In this technique, "DNA enters the plant cells through miniature pores which are temporarily caused by electric pulses."

Microinjection directly injects the gene into the DNA.

Plant scientists, backed by results of modern comprehensive profiling of crop composition, point out that crops modified using GM techniques are less likely to have unintended changes than are conventionally bred crops.

In research tobacco and *Arabidopsis thaliana* are the most frequently modified plants, due to well-developed transformation methods, easy propagation and well studied genomes. They serve as model organisms for other plant species.

Introducing new genes into plants requires a promoter specific to the area where the gene is to be expressed. For instance, to express a gene only in rice grains and not in leaves, an endosperm-specific promoter is used. The codons of the gene must be optimized for the organism due to codon usage bias.

Types of Modifications

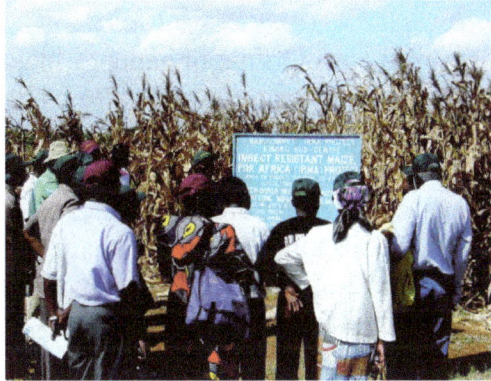

Transgenic maize containing a gene from the bacteria Bacillus thuringiensis

Transgenic

Transgenic plants have genes inserted into them that are derived from another species. The inserted genes can come from species within the same kingdom (plant to plant) or between kingdoms (for example, bacteria to plant). In many cases the inserted DNA has to be modified slightly in order to correctly and efficiently express in the host organism. Transgenic plants are used to express proteins like the cry toxins from *B. thuringiensis*, herbicide resistant genes, antibodies and antigens for vaccinations A study led by the European Food Safety Authority (EFSA) found also viral genes in transgenic plants.

Transgenic carrots have been used to produce the drug Taliglucerase alfa which is used to treat Gaucher's disease. In the laboratory, transgenic plants have been modified to increase photosynthesis (currently about 2% at most plants versus the theoretic potential of 9–10%). This is possible by changing the rubisco enzyme (i.e. changing C3 plants into C4 plants), by placing the rubisco in a carboxysome, by adding CO_2 pumps in the cell wall, by changing the leaf form/size. Plants have been engineered to exhibit bioluminescence that may become a sustainable alternative to electric lighting.

Cisgenic

Cisgenic plants are made using genes found within the same species or a closely related one, where conventional plant breeding can occur. Some breeders and scientists argue that cisgenic modification is useful for plants that are difficult to crossbreed by conventional means (such as potatoes), and that plants in the cisgenic category should not require the same regulatory scrutiny as transgenics.

Subgenic

Genetically modified plants can also be developed using gene knockdown to alter the genetic makeup of a plant without incorporating genes from other plants. In 2014, Chinese researcher Gao Caixia filed patents on the creation of a strain of wheat that is resistant to powdery mildew. The strain lacks genes that encode proteins that repress defenses against the mildew. The researchers deleted all three copies of the genes from wheat's hexaploid genome. Gao used the TALENs and CRISPR gene editing tools without adding or changing any other genes. No field trials were immediately planned. The CRISPR technique has also been used to modify white button mushrooms (*Agaricus bisporus*).

Economics

GM food's economic value to farmers is one of its major benefits, including in developing nations. A 2010 study found that Bt corn provided economic benefits of $6.9 billion over the previous 14 years in five Midwestern states. The majority ($4.3 billion) accrued to farmers producing non-Bt corn. This was attributed to European corn borer populations reduced by exposure to Bt corn, leaving fewer to attack conventional corn nearby. Agriculture economists calculated that "world surplus [increased by] $240.3 million for 1996. Of this total, the largest share (59%) went to U.S. farmers. Seed company Monsanto received the next largest share (21%), followed by US consumers (9%), the rest of the world (6%), and the germplasm supplier, Delta & Pine Land Company of Mississippi (5%)."

According to the International Service for the Acquisition of Agri-biotech Applications (ISAAA), in 2014 approximately 18 million farmers grew biotech crops in 28 countries; about 94% of the farmers were resource-poor in developing countries. 53% of the global biotech crop area of 181.5 million hectares was grown in 20 developing countries. PG Economics comprehensive 2012 study concluded that GM crops increased farm incomes worldwide by $14 billion in 2010, with over half this total going to farmers in developing countries.

Critics challenged the claimed benefits to farmers over the prevalence of biased observers and by the absence of randomized controlled trials. The main Bt crop grown by small farmers in developing countries is cotton. A 2006 review of Bt cotton findings by agricultural economists concluded, "the overall balance sheet, though promising, is mixed. Economic returns are highly variable over years, farm type, and geographical location".

In 2013 the European Academies Science Advisory Council (EASAC) asked the EU to allow the development of agricultural GM technologies to enable more sustainable agriculture, by employing fewer land, water and nutrient resources. EASAC also criticizes the EU's "timeconsuming and expensive regulatory framework" and said that the EU had fallen behind in the adoption of GM technologies.

Participants in agriculture business markets include seed companies, agrochemical companies, distributors, farmers, grain elevators and universities that develop new crops/traits and whose agricultural extensions advise farmers on best practices. According to a 2012 review based on data from the late 1990s and early 2000s, much of the GM crop grown each year is used for livestock feed and increased demand for meat leads to increased demand for GM feedcrops. Feed grain usage as a percentage of total crop production is 70% for corn and more than 90% of oil seed meals such as soybeans. About 65 million metric tons of GM corn grains and about 70 million metric tons of soybean meals derived from GM soybean become feed.

In 2014 the global value of biotech seed was US$15.7 billion; US$11.3 billion (72%) was in industrial countries and US$4.4 billion (28%) was in the developing countries. In 2009, Monsanto had $7.3 billion in sales of seeds and from licensing its technology; DuPont, through its Pioneer subsidiary, was the next biggest company in that market. As of 2009, the overall Roundup line of products including the GM seeds represented about 50% of Monsanto's business.

Some patents on GM traits have expired, allowing the legal development of generic strains that include these traits. For example, generic glyphosate-tolerant GM soybean is now available. Another impact is that traits developed by one vendor can be added to another vendor's proprietary strains, potentially increasing product choice and competition. The patent on the first type of *Roundup Ready* crop that Monsanto produced (soybeans) expired in 2014 and the first harvest of off-patent soybeans occurs in the spring of 2015. Monsanto has broadly licensed the patent to other seed companies that include the glyphosate resistance trait in their seed products. About 150 companies have licensed the technology, including Syngenta and DuPont Pioneer.

Yield

In 2014, the largest review yet concluded that GM crops' effects on farming were positive. The meta-analysis considered all published English-language examinations of the agronomic and economic impacts between 1995 and March 2014 for three major GM crops: soybean, maize, and cotton. The study found that herbicide-tolerant crops have lower production costs, while for insect-resistant crops the reduced pesticide use was offset by higher seed prices, leaving overall production costs about the same.

Yields increased 9% for herbicide tolerance and 25% for insect resistant varieties. Farmers who adopted GM crops made 69% higher profits than those who did not. The review found that GM crops help farmers in developing countries, increasing yields by 14 percentage points.

The researchers considered some studies that were not peer-reviewed, and a few that did not report sample sizes. They attempted to correct for publication bias, by considering sources beyond academic journals. The large data set allowed the study to control for potentially confounding variables such as fertiliser use. Separately, they concluded that the funding source did not influence study results.

Traits

GM crops grown today, or under development, have been modified with various traits. These traits include improved shelf life, disease resistance, stress resistance, herbicide resistance, pest resistance, production of useful goods such as biofuel or drugs, and ability to absorb toxins and for use in bioremediation of pollution.

Recently, research and development has been targeted to enhancement of crops that are locally important in developing countries, such as insect-resistant cowpea for Africa and insect-resistant brinjal (eggplant).

Lifetime

The first genetically modified crop approved for sale in the U.S. was the *FlavrSavr* tomato, which had a longer shelf life. It is no longer on the market.

In November 2014, the USDA approved a GM potato that prevents bruising.

In February 2015 Arctic Apples were approved by the USDA, becoming the first genetically modified apple approved for US sale. Gene silencing was used to reduce the expression of polyphenol oxidase (PPO), thus preventing enzymatic browning of the fruit after it has been sliced open. The trait was added to Granny Smith and Golden Delicious varieties. The trait includes a bacterial antibiotic resistance gene that provides resistance to the antibiotic kanamycin. The genetic engineering involved cultivation in the presence of kanamycin, which allowed only resistant cultivars to survive. Humans consuming apples do not acquire kanamycin resistance, per arcticapple.com. The FDA approved the apples in March 2015.

Nutrition

Edible Oils

Some GM soybeans offer improved oil profiles for processing or healthier eating. Camelina sativa has been modified to produce plants that accumulate high levels of oils similar to fish oils.

Vitamin Enrichment

Golden rice, developed by the International Rice Research Institute (IRRI), provides greater amounts of Vitamin A targeted at reducing Vitamin A deficiency. As of January 2016, golden rice has not yet been grown commercially in any country.

Researchers vitamin-enriched corn derived from South African white corn variety M37W, producing a 169-fold increase in Vitamin A, 6-fold increase in Vitamin C and doubled concentrations of folate. Modified Cavendish bananas express 10-fold the amount of Vitamin A as unmodified varieties.

Toxin Reduction

A genetically modified cassava under development offers lower cyanogen glucosides and enhanced protein and other nutrients (called BioCassava).

In November 2014, the USDA approved a potato, developed by J.R. Simplot Company, that prevents bruising and produces less acrylamide when fried. The modifications prevent natural, harmful proteins from being made via RNA interference. They do not employ genes from non-potato species. The trait was added to the Russet Burbank, Ranger Russet and Atlantic varieties.

Stress Resistance

Plants engineered to tolerate non-biological stressors such as drought, frost, high soil salinity, and nitrogen starvation were in development. In 2011, Monsanto's DroughtGard maize became the first drought-resistant GM crop to receive US marketing approval.

Herbicides

Glyphosate

As of 1999 the most prevalent GM trait was glyphosate-resistance. Glyphosate, (the active ingredient in Roundup and other herbicide products) kills plants by interfering with the shikimate pathway in plants, which is essential for the synthesis of the aromatic amino acids phenylalanine, tyrosine and tryptophan. The shikimate pathway is not present in animals, which instead obtain aromatic amino acids from their diet. More specifically, glyphosate inhibits the enzyme 5-enolpyruvylshikimate-3-phosphate synthase (EPSPS).

This trait was developed because the herbicides used on grain and grass crops at the time were highly toxic and not effective against narrow-leaved weeds. Thus, developing crops that could withstand spraying with glyphosate would both reduce environmental and health risks, and give an agricultural edge to the farmer.

Some micro-organisms have a version of EPSPS that is resistant to glyphosate inhibition. One of these was isolated from an *Agrobacterium* strain CP4 (CP4 EPSPS) that was resistant to glyphosate. The CP4 EPSPS gene was engineered for plant expression by fusing the 5' end of the gene to a chloroplast transit peptide derived from the petunia EPSPS. This transit peptide was used because it had shown previously an ability to deliver bacterial EPSPS to the chloroplasts of other plants. This CP4 EPSPS gene was cloned and transfected into soybeans.

The plasmid used to move the gene into soybeans was PV-GMGTO4. It contained three bacterial genes, two CP4 EPSPS genes, and a gene encoding beta-glucuronidase (GUS) from *Escherichia coli* as a marker. The DNA was injected into the soybeans using the particle acceleration method. Soybean cultivar A5403 was used for the transformation.

Bromoxynil

Tobacco plants have been engineered to be resistant to the herbicide bromoxynil.

Glufosinate

Crops have been commercialized that are resistant to the herbicide glufosinate, as well. Crops engineered for resistance to multiple herbicides to allow farmers to use a mixed group of two, three, or four different chemicals are under development to combat growing herbicide resistance.

2,4-D

In October 2014 the US EPA registered Dow's Enlist Duo maize, which is genetically modified to be resistant to both glyphosate and 2,4-D, in six states. Inserting a bacterial aryloxyalkanoate dioxygenase gene, *aad1* makes the corn resistant to 2,4-D. The USDA had approved maize and soybeans with the mutation in September 2014.

Dicamba

Monsanto has requested approval for a stacked strain that is tolerant of both glyphosate and dicamba.

Pest Resistance

Insects

Tobacco, corn, rice and many other crops have been engineered to express genes encoding for insecticidal proteins from Bacillus thuringiensis (Bt). Papaya, potatoes, and squash have been engineered to resist viral pathogens such as cucumber mosaic virus which, despite its name, infects a wide variety of plants. The introduction of Bt crops during the period between 1996 and 2005 has been estimated to have reduced the total volume of insecticide active ingredient use in the United States by over 100 thousand tons. This represents a 19.4% reduction in insecticide use.

In the late 1990s, a genetically modified potato that was resistant to the Colorado potato beetle was withdrawn because major buyers rejected it, fearing consumer opposition.

Viruses

Virus resistant papaya were developed In response to a papaya ringspot virus (PRV) outbreak in Hawaii in the late 1990s. . They incorporate PRV DNA. By 2010, 80% of Hawaiian papaya plants were genetically modified.

Potatoes were engineered for resistance to potato leaf roll virus and Potato virus Y in 1998. Poor sales led to their market withdrawal after three years.

Yellow squash that were resistant to at first two, then three viruses were developed, beginning in the 1990s. The viruses are watermelon, cucumber and zucchini/courgette yellow mosaic. Squash was the second GM crop to be approved by US regulators. The trait was later added to zucchini.

Many strains of corn have been developed in recent years to combat the spread of Maize dwarf mosaic virus, a costly virus that causes stunted growth which is carried in Johnson grass and spread by aphid insect vectors. These strands are commercially available although the resistance is not standard among GM corn variants.

By-products

Drugs

In 2012, the FDA approved the first plant-produced pharmaceutical, a treatment for Gaucher's Disease. Tobacco plants have been modified to produce therapeutic antibodies.

Biofuel

Algae is under development for use in biofuels. Researchers in Singapore were working on GM jatropha for biofuel production. Syngenta has USDA approval to market a maize trademarked Enogen that has been genetically modified to convert its starch to sugar for ethanol. In 2013, the Flemish Institute for Biotechnology was investigating poplar trees genetically engineered to contain less lignin to ease conversion into ethanol. Lignin is the critical limiting factor when using wood to make bio-ethanol because lignin limits the accessibility of cellulose microfibrils to depolymerization by enzymes.

Materials

Companies and labs are working on plants that can be used to make bioplastics. Potatoes that produce industrially useful starches have been developed as well. Oilseed can be modified to produce fatty acids for detergents, substitute fuels and petrochemicals.

Bioremediation

Scientists at the University of York developed a weed (*Arabidopsis thaliana*) that contains genes from bacteria that can clean TNT and RDX-explosive soil contaminants. 16 million hectares in the USA (1.5% of the total surface) are estimated to be contaminated with TNT and RDX. However *A. thaliana* was not tough enough for use on military test grounds.

Genetically modified plants have been used for bioremediation of contaminated soils. Mercury, selenium and organic pollutants such as polychlorinated biphenyls (PCBs).

Marine environments are especially vulnerable since pollution such as oil spills are not containable. In addition to anthropogenic pollution, millions of tons of petroleum annually enter the marine environment from natural seepages. Despite its toxicity, a considerable fraction of petroleum oil entering marine systems is eliminated by the hydrocarbon-degrading activities of microbial communities. Particularly successful is a recently discovered group of specialists, the so-called hydrocarbonoclastic bacteria (HCCB) that may offer useful genes.

Asexual Reproduction

Crops such as maize reproduce sexually each year. This randomizes which genes get propagated to the next generation, meaning that desirable traits can be lost. To maintain a high-quality crop, some farmers purchase seeds every year. Typically, the seed company maintains two inbred varieties, and crosses them into a hybrid strain that is then sold. Related plants like sorghum and gamma grass are able to perform apomixis, a form of asexual reproduction that keeps the plant's DNA intact. This trait is apparently controlled by a single dominant gene, but traditional breeding has been unsuccessful in creating asexually-reproducing maize. Genetic engineering offers another route to this goal. Successful modification would allow farmers to replant harvested seeds that retain desirable traits, rather than relying on purchased seed.

Crops

Herbicide tolerance

GMO	Use	Countries approved in	First approved	Notes
Alfalfa	Animal feed	USA	2005	Approval withdrawn in 2007 and then re-approved in 2011
Canola	Cooking oil Margarine Emulsifiers in packaged foods	Australia	2003	
		Canada	1995	
		USA	1995	

		Argentina	2001	
		Australia	2002	
		Brazil	2008	
		Columbia	2004	
Cotton	Fiber Cottonseed oil Animal feed	Costa Rica	2008	
		Mexico	2000	
		Paraguay	2013	
		South Africa	2000	
		USA	1994	
		Argentina	1998	
		Brazil	2007	
		Canada	1996	
		Colombia	2007	
		Cuba	2011	
Maize	Animal feed high-fructose corn syrup corn starch	European Union	1998	Grown in Portugal, Spain, Czech Republic, Slovakia and Romania
		Honduras	2001	
		Paraguay	2012	
		Philippines	2002	
		South Africa	2002	
		USA	1995	
		Uruguay	2003	
		Argentina	1996	
		Bolivia	2005	
		Brazil	1998	
		Canada	1995	
		Chile	2007	
Soybean	Animal feed Soybean oil	Costa Rica	2001	
		Mexico	1996	
		Paraguay	2004	
		South Africa	2001	
		USA	1993	
		Uruguay	1996	
Sugar Beet	Food	Canada	2001	
		USA	1998	Commercialised 2007, production blocked 2010, resumed 2011.

Insect Resistance

GMO	Use	Countries approved in	First approved	Notes
Cotton	Fiber Cottonseed oil Animal feed	Argentina	1998	
		Australia	2003	
		Brazil	2005	
		Burkina Faso	2009	
		China	1997	
		Colombia	2003	
		Costa Rica	2008	
		India	2002	Largest producer of Bt cotton
		Mexico	1996	
		Myanmar	2006[N 1]	
		Pakistan	2010[N 1]	
		Paraguay	2007	
		South Africa	1997	
		Sudan	2012	
		USA	1995	
Eggplant	Food	Bangladesh	2013	12 ha planted on 120 farms in 2014
Maize	Animal feed high-fructose corn syrup corn starch	Argentina	1998	
		Brazil	2005	
		Columbia	2003	
		Mexico	1996	Centre of origin for maize
		Paraguay	2007	
		Philippines	2002	
		South Africa	1997	
		Uruguay	2003	
		USA	1995	
Poplar	Tree	China	1998	543 ha of bt poplar planted in 2014

Other Modified Traits

GMO	Use	Trait	Countries approved in	First approved	Notes
Canola	Cooking oil Margarine Emulsifiers in packaged foods	High laurate canola	Canada	1996	
			USA	1994	
		Phytase production	USA	1998	

			Australia	1995	
Carnation	Ornamental	Delayed senescence	Australia	1995	
			Norway	1998	
		Modified flower colour	Australia	1995	
			Columbia	2000	In 2014 4 ha were grown in greenhouses for export
			European Union	1998	Two events expired 2008, another approved 2007
			Japan	2004	
			Malaysia	2012	For ornamental purposes
			Norway	1997	
Maize	Animal feed high-fructose corn syrup corn starch	Increased lysine	Canada	2006	
			USA	2006	
		Drought tolerance	Canada	2010	
			USA	2011	
Papaya	Food	Virus resistance	China	2006	
			USA	1996	Mostly grown in Hawaii
Petunia	Ornamental	Modified flower colour	China	1997	
Potato	Food	Virus resistance	Canada	1999	
			USA	1997	
	Industrial	Modified starch	USA	2014	
Rose	Ornamental	Modified flower colour	Australia	2009	Surrendered renewal
			Colombia	2010[N 2]	Greenhouse cultivation for export only.
			Japan	2008	
			USA	2011	
Soybean	Animal feed Soybean oil	Increased oleic acid production	Argentina	2015	
			Canada	2000	
			USA	1997	
		Stearidonic acid production	Canada	2011	
			USA	2011	
Squash	Food	Virus resistance	USA	1994	
Sugar Cane	Food	Drought tolerance	Indonesia	2013	Environmental certificate only
Tobacco	Cigarettes	Nicotine reduction	USA	2002	

Development

The number of USDA-approved field releases for testing grew from 4 in 1985 to 1,194 in 2002 and averaged around 800 per year thereafter. The number of sites per release and the number of gene constructs (ways that the gene of interest is packaged together with other elements)—have rapidly increased since 2005. Releases with agronomic properties (such as drought resistance) jumped from 1,043 in 2005 to 5,190 in 2013. As of September 2013, about 7,800 releases had been approved for corn, more than 2,200 for soybeans, more than 1,100 for cotton, and about 900 for potatoes. Releases were approved for herbicide tolerance (6,772 releases), insect resistance (4,809), product quality such as flavor or nutrition (4,896), agronomic properties like drought resistance (5,190), and virus/fungal resistance (2,616). The institutions with the most authorized field releases include Monsanto with 6,782, Pioneer/DuPont with 1,405, Syngenta with 565, and USDA's Agricultural Research Service with 370. As of September 2013 USDA had received proposals for releasing GM rice, squash, plum, rose, tobacco, flax and chicory.

Farming practices

Bt Resistance

Constant exposure to a toxin creates evolutionary pressure for pests resistant to that toxin. Overreliance on glyphosate and a reduction in the diversity of weed management practices allowed the spread of glyphosate resistance in 14 weed species/biotypes in the US.

One method of reducing resistance is the creation of refuges to allow nonresistant organisms to survive and maintain a susceptible population.

To reduce resistance to Bt crops, the 1996 commercialization of transgenic cotton and maize came with a management strategy to prevent insects from becoming resistant. Insect resistance management plans are mandatory for Bt crops. The aim is to encourage a large population of pests so that any (recessive) resistance genes are diluted within the population. Resistance lowers evolutionary fitness in the absence of the stressor (Bt). In refuges, non-resistant strains outcompete resistant ones.

With sufficiently high levels of transgene expression, nearly all of the heterozygotes (S/s), i.e., the largest segment of the pest population carrying a resistance allele, will be killed before maturation, thus preventing transmission of the resistance gene to their progeny. Refuges (i. e., fields of non-transgenic plants) adjacent to transgenic fields increases the likelihood that homozygous resistant (s/s) individuals and any surviving heterozygotes will mate with susceptible (S/S) individuals from the refuge, instead of with other individuals carrying the resistance allele. As a result, the resistance gene frequency in the population remains lower.

Complicating factors can affect the success of the high-dose/refuge strategy. For example, if the temperature is not ideal, thermal stress can lower Bt toxin production and leave the plant more susceptible. More importantly, reduced late-season expression has been documented, possibly resulting from DNA methylation of the promoter. The success of the high-dose/refuge strategy has successfully maintained the value of Bt crops. This success has depended on factors independent of management strategy, including low initial resistance allele frequencies, fitness costs associated with resistance, and the abundance of non-Bt host plants outside the refuges.

Companies that produce Bt seed are introducing strains with multiple Bt proteins. Monsanto did this with Bt cotton in India, where the product was rapidly adopted. Monsanto has also; in an attempt to simplify the process of implementing refuges in fields to comply with Insect Resistance Management(IRM) policies and prevent irresponsible planting practices; begun marketing seed bags with a set proportion of refuge (non-transgenic) seeds mixed in with the Bt seeds being sold. Coined "Refuge-In-a-Bag" (RIB), this practice is intended to increase farmer compliance with refuge requirements and reduce additional labor needed at planting from having separate Bt and refuge seed bags on hand. This strategy is likely to reduce the likelihood of Bt-resistance occurring for corn rootworm, but may increase the risk of resistance for lepidopteran corn pests, such as European corn borer. Increased concerns for resistance with seed mixtures include partially resistant larvae on a Bt plant being able to move to a susceptible plant to survive or cross pollination of refuge pollen on to Bt plants that can lower the amount of Bt expressed in kernels for ear feeding insects.

Herbicide Resistance

Best management practices (BMPs) to control weeds may help delay resistance. BMPs include applying multiple herbicides with different modes of action, rotating crops, planting weed-free seed, scouting fields routinely, cleaning equipment to reduce the transmission of weeds to other fields, and maintaining field borders. The most widely planted GMOs are designed to tolerate herbicides. By 2006 some weed populations had evolved to tolerate some of the same herbicides. Palmer amaranth is a weed that competes with cotton. A native of the southwestern US, it traveled east and was first found resistant to glyphosate in 2006, less than 10 years after GM cotton was introduced.

Plant Protection

Farmers generally use less insecticide when they plant Bt-resistant crops. Insecticide use on corn farms declined from 0.21 pound per planted acre in 1995 to 0.02 pound in 2010. This is consistent with the decline in European corn borer populations as a direct result of Bt corn and cotton. The establishment of minimum refuge requirements helped delay the evolution of Bt resistance. However resistance appears to be developing to some Bt traits in some areas.

Tillage

By leaving at least 30% of crop residue on the soil surface from harvest through planting, conservation tillage reduces soil erosion from wind and water, increases water retention, and reduces soil degradation as well as water and chemical runoff. In addition, conservation tillage reduces the carbon footprint of agriculture. A 2014 review covering 12 states from 1996 to 2006, found that a 1% increase in herbicde-tolerant (HT) soybean adoption leads to a 0.21% increase in conservation tillage and a 0.3% decrease in quality-adjusted herbicide use.

Regulation

The regulation of genetic engineering concerns the approaches taken by governments to assess and manage the risks associated with the development and release of genetically modified crops. There are differences in the regulation of GM crops between countries, with some of the most marked differences occurring between the USA and Europe. Regulation varies in a given country depending on the intended use of each product. For example, a crop not intended for food use is generally not reviewed by authorities responsible for food safety.

Production

In 2013, GM crops were planted in 27 countries; 19 were developing countries and 8 were developed countries. 2013 was the second year in which developing countries grew a majority (54%) of the total GM harvest. 18 million farmers grew GM crops; around 90% were small-holding farmers in developing countries.

Country	2013– GM planted area (million hectares)	Biotech crops
USA	70.1	Maize, Soybean, Cotton, Canola, Sugarbeet, Alfalfa, Papaya, Squash
Brazil	40.3	Soybean, Maize, Cotton
Argentina	24.4	Soybean, Maize, Cotton
India	11.0	Cotton
Canada	10.8	Canola, Maize, Soybean, Sugarbeet
Total	175.2	----

The United States Department of Agriculture (USDA) reports every year on the total area of GMO varieties planted in the United States. According to National Agricultural Statistics Service, the states published in these tables represent 81–86 percent of all corn planted area, 88–90 percent of all soybean planted area, and 81–93 percent of all upland cotton planted area (depending on the year).

Global estimates are produced by the International Service for the Acquisition of Agri-biotech Applications (ISAAA) and can be found in their annual reports, "Global Status of Commercialized Transgenic Crops".

Farmers have widely adopted GM technology. Between 1996 and 2013, the total surface area of land cultivated with GM crops increased by a factor of 100, from 17,000 square kilometers (4,200,000 acres) to 1,750,000 km^2 (432 million acres). 10% of the world's arable land was planted with GM crops in 2010. As of 2011, 11 different transgenic crops were grown commercially on 395 million acres (160 million hectares) in 29 countries such as the USA, Brazil, Argentina, India, Canada, China, Paraguay, Pakistan, South Africa, Uruguay, Bolivia, Australia, Philippines, Myanmar, Burkina Faso, Mexico and Spain. One of the key reasons for this widespread adoption is the perceived economic benefit the technology brings to farmers. For example, the system of planting glyphosate-resistant seed and then applying glyphosate once plants emerged provided farmers with the opportunity to dramatically increase the yield from a given plot of land, since this allowed them to plant rows closer together. Without it, farmers had to plant rows far enough apart to control post-emergent weeds with mechanical tillage. Likewise, using Bt seeds means that farmers do not have to purchase insecticides, and then invest time, fuel, and equipment in applying them. However critics have disputed whether yields are higher and whether chemical use is less, with GM crops.

In the US, by 2014, 94% of the planted area of soybeans, 96% of cotton and 93% of corn were genetically modified varieties. Genetically modified soybeans carried herbicide-tolerant traits only, but maize and cotton carried both herbicide tolerance and insect protection traits (the latter largely Bt protein). These constitute "input-traits" that are aimed to financially benefit the producers, but may have indirect environmental benefits and cost benefits to consumers. The Grocery Manufacturers of America estimated in 2003 that 70–75% of all processed foods in the U.S. contained a GM ingredient.

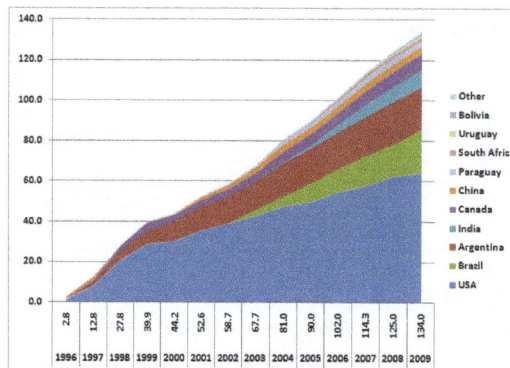

Land area used for genetically modified crops by country (1996–2009), in millions of hectares. In 2011, the land area used was 160 million hectares, or 1.6 million square kilometers.

Europe grows relatively few genetically engineered crops with the exception of Spain, where one fifth of maize is genetically engineered, and smaller amounts in five other countries. The EU had a 'de facto' ban on the approval of new GM crops, from 1999 until 2004. GM crops are now regulated by the EU. In 2015, genetically engineered crops are banned in 38 countries worldwide, 19 of them in Europe. Developing countries grew 54 percent of genetically engineered crops in 2013.

In recent years GM crops expanded rapidly in developing countries. In 2013 approximately 18 million farmers grew 54% of worldwide GM crops in developing countries. 2013's largest increase was in Brazil (403,000 km² versus 368,000 km² in 2012). GM cotton began growing in India in 2002, reaching 110,000 km² in 2013.

According to the 2013 ISAAA brief: "…a total of 36 countries (35 + EU-28) have granted regulatory approvals for biotech crops for food and/or feed use and for environmental release or planting since 1994… a total of 2,833 regulatory approvals involving 27 GM crops and 336 GM events (NB: an "event" is a specific genetic modification in a specific species) have been issued by authorities, of which 1,321 are for food use (direct use or processing), 918 for feed use (direct use or processing) and 599 for environmental release or planting. Japan has the largest number (198), followed by the U.S.A. (165, not including "stacked" events), Canada (146), Mexico (131), South Korea (103), Australia (93), New Zealand (83), European Union (71 including approvals that have expired or under renewal process), Philippines (68), Taiwan (65), Colombia (59), China (55) and South Africa (52). Maize has the largest number (130 events in 27 countries), followed by cotton (49 events in 22 countries), potato (31 events in 10 countries), canola (30 events in 12 countries) and soybean (27 events in 26 countries).

Controversy

GM foods are controversial and the subject of protests, vandalism, referenda, legislation, court action and scientific disputes. The controversies involve consumers, biotechnology companies, governmental regulators, non-governmental organizations and scientists. The key areas are whether GM food should be labeled, the role of government regulators, the effect of GM crops on health and the environment, the effects of pesticide use and resistance, the impact on farmers, and their roles in feeding the world and energy production.

There is a scientific consensus that currently available food derived from GM crops poses no greater risk to human health than conventional food, but that each GM food needs to be tested on a case-by-case basis before introduction. Nonetheless, members of the public are much less likely than scientists to perceive GM foods as safe. The legal and regulatory status of GM foods varies by country, with some nations banning or restricting them, and others permitting them with widely differing degrees of regulation.

No reports of ill effects have been documented in the human population from GM food. Although GMO labeling is required in many countries, the United States Food and Drug Administration does not require labeling, nor does it recognize a distinction between approved GMO and non-GMO foods.

Advocacy groups such as Center for Food Safety, Union of Concerned Scientists, Greenpeace and the World Wildlife Fund claim that risks related to GM food have not been adequately examined and managed, that GMOs are not sufficiently tested and should be labelled, and that regulatory authorities and scientific bodies are too closely tied to industry. Some studies have claimed that genetically modified crops can cause harm; a 2016 review that reanalyzed the data from six of these studies found that their statistical methodologies were flawed and did not demonstrate harm, and said that conclusions about GMO crop safety should be drawn from "the totality of the evidence... instead of far-fetched evidence from single studies".

Genetically Modified Food

Genetically modified foods or GM foods, also genetically engineered foods, are foods produced from organisms that have had changes introduced into their DNA using the methods of genetic engineering. Genetic engineering techniques allow for the introduction of new traits as well as greater control over traits than previous methods such as selective breeding and mutation breeding.

Commercial sale of genetically modified foods began in 1994, when Calgene first marketed its unsuccessful Flavr Savr delayed-ripening tomato. Most food modifications have primarily focused on cash crops in high demand by farmers such as soybean, corn, canola, and cotton. Genetically modified crops have been engineered for resistance to pathogens and herbicides and for better nutrient profiles. GM livestock have been developed, although as of November 2013 none were on the market.

There is a scientific consensus that currently available food derived from GM crops poses no greater risk to human health than conventional food, but that each GM food needs to be tested on a case-by-case basis before introduction. Nonetheless, members of the public are much less likely than scientists to perceive GM foods as safe. The legal and regulatory status of GM foods varies by country, with some nations banning or restricting them, and others permitting them with widely differing degrees of regulation.

However, there are ongoing public concerns related to food safety, regulation, labelling, environmental impact, research methods, and the fact that some GM seeds are subject to intellectual property rights owned by corporations.

Definition

Genetically modified foods, GM foods or genetically engineered foods, are foods produced from organisms that have had changes introduced into their DNA using the methods of genetic engineering as opposed to traditional cross breeding. In the US, the Department of Agriculture (USDA) and the Food and Drug Administration (FDA) favor the use of "genetic engineering" over "genetic modification" as the more precise term; the USDA defines genetic modification to include "genetic engineering or other more traditional methods."

According to the World Health Organization, "Genetically modified organisms (GMOs) can be defined as organisms (i.e. plants, animals or microorganisms) in which the genetic material (DNA) has been altered in a way that does not occur naturally by mating and/or natural recombination. The technology is often called 'modern biotechnology' or 'gene technology', sometimes also 'recombinant DNA technology' or 'genetic engineering'. ... Foods produced from or using GM organisms are often referred to as GM foods."

History

Human-directed genetic manipulation of food began with the domestication of plants and animals through artificial selection at about 10,500 to 10,100 BC.[1] The process of selective breeding, in which organisms with desired traits (and thus with the desired genes) are used to breed the next generation and organisms lacking the trait are not bred, is a precursor to the modern concept of genetic modification (GM). With the discovery of DNA in the early 1900s and various advancements in genetic techniques through the 1970s it became possible to directly alter the DNA and genes within food.

Genetically modified microbial enzymes were the first application of genetically modified organisms in food production and were approved in 1988 by the US Food and Drug Administration. In the early 1990s, recombinant chymosin was approved for use in several countries. Cheese had typically been made using the enzyme complex rennet that had been extracted from cows' stomach lining. Scientists modified bacteria to produce chymosin, which was also able to clot milk, resulting in cheese curds.

The first genetically modified food approved for release was the Flavr Savr tomato in 1994. Developed by Calgene, it was engineered to have a longer shelf life by inserting an antisense gene that delayed ripening. In 1995, *Bacillus thuringiensis* (Bt) Potato was approved for cultivation, making it the first pesticide producing crop to be approved in the USA. Other genetically modified crops receiving marketing approval in 1995 were: canola with modified oil composition, Bt maize, cotton resistant to the herbicide bromoxynil, Bt cotton, glyphosate-tolerant soybeans, virus-resistant squash, and another delayed ripening tomato.

With the creation of golden rice in 2000, scientists had genetically modified food to increase its nutrient value for the first time.

By 2010, 29 countries had planted commercialized biotech crops and a further 31 countries had granted regulatory approval for transgenic crops to be imported. The US was the leading country in the production of GM foods in 2011, with twenty-five GM crops having received regulatory approval. In 2015, 92% of corn, 94% of soybeans, and 94% of cotton produced in the US were genetically modified strains.

The first genetically modified animal to be approved for food use was AquAdvantage salmon in 2015. The salmon were transformed with a growth hormone-regulating gene from a Pacific Chinook salmon and a promoter from an ocean pout enabling it to grow year-round instead of only during spring and summer.

In April 2016, a white button mushroom (*Agaricus bisporus*) modified using the CRISPR technique received *de facto* approval in the United States, after the USDA said it would not have to go through the agency's regulatory process. The agency considers the mushroom exempt because the editing process did not involve the introduction of foreign DNA.

The most widely planted GMOs are designed to tolerate herbicides. By 2006 some weed populations had evolved to tolerate some of the same herbicides. Palmer amaranth is a weed that competes with cotton. A native of the southwestern US, it traveled east and was first found resistant to glyphosate in 2006, less than 10 years after GM cotton was introduced.

Process

Genetically engineered organisms are generated and tested in the laboratory for desired qualities. The most common modification is to add one or more genes to an organism's genome. Less commonly, genes are removed or their expression is increased or silenced or the number of copies of a gene is increased or decreased.

Once satisfactory strains are produced, the producer applies for regulatory approval to field-test them, called a "field release." Field-testing involves cultivating the plants on farm fields or growing animals in a controlled environment. If these field tests are successful, the producer applies for regulatory approval to grow and market the crop. Once approved, specimens (seeds, cuttings, breeding pairs, etc.) are cultivated and sold to farmers. The farmers cultivate and market the new strain. In some cases, the approval covers marketing but not cultivation.

According to the USDA, the number of field releases for genetically engineered organisms has grown from four in 1985 to an average of about 800 per year. Cumulatively, more than 17,000 releases had been approved through September 2013.

Crops

Fruits and Vegetables

Papaya was genetically modified to resist the ringspot virus. 'SunUp' is a transgenic red-fleshed Sunset papaya cultivar that is homozygous for the coat protein gene PRSV; 'Rainbow' is a yellow-fleshed F1 hybrid developed by crossing 'SunUp' and nontransgenic yellow-fleshed 'Kapoho'. The New York Times stated, "in the early 1990s, Hawaii's papaya industry was facing disaster because of the deadly papaya ringspot virus. Its single-handed savior was a breed engineered to be resistant to the virus. Without it, the state's papaya industry would have collapsed. Today, 80% of Hawaiian papaya is genetically engineered, and there is still no conventional or organic method to control ringspot virus." The GM cultivar was approved in 1998. In China, a transgenic PRSV-resistant papaya was developed by South China Agricultural University and was first approved for commercial planting in 2006; as of 2012 95% of the papaya grown in Guangdong province and 40% of the papaya grown in Hainan province was genetically modified.

3 views of the Sunset papaya cultivar, which was genetically modified to create the SunUp cultivar, resistant to PRSV.

The New Leaf potato, brought to market by Monsanto in the late 1990s, was developed for the fast food market. It was withdrawn in 2001 after retailers rejected it and food processors ran into export problems.

As of 2005, about 13% of the Zucchini (a form of squash) grown in the US was genetically modified to resist three viruses; that strain is also grown in Canada.

Plums genetically engineered for resistance to plum pox, a disease carried by aphids.

In 2011, BASF requested the European Food Safety Authority's approval for cultivation and marketing of its Fortuna potato as feed and food. The potato was made resistant to late blight by adding resistant genes blb1 and blb2 that originate from the Mexican wild potato Solanum bulbocastanum. In February 2013, BASF withdrew its application.

In 2013, the USDA approved the import of a GM pineapple that is pink in color and that "over-expresses" a gene derived from tangerines and suppress other genes, increasing production of lycopene. The plant's flowering cycle was changed to provide for more uniform growth and quality. The fruit "does not have the ability to propagate and persist in the environment once they have been harvested," according to USDA APHIS. According to Del Monte's submission, the pineapples are commercially grown in a "monoculture" that prevents seed production, as the plant's flowers aren't exposed to compatible pollen sources. Importation into Hawaii is banned for "plant sanitation" reasons.

In 2014, the USDA approved a genetically modified potato developed by J.R. Simplot Company that contained ten genetic modifications that prevent bruising and produce less acrylamide when fried. The modifications eliminate specific proteins from the potatoes, via RNA interference, rather than introducing novel proteins.

In February 2015 Arctic Apples were approved by the USDA, becoming the first genetically modified apple approved for sale in the US. Gene silencing is used to reduce the expression of polyphenol oxidase (PPO), thus preventing the fruit from browning.

Corn

Corn used for food and ethanol has been genetically modified to tolerate various herbicides and to express a protein from Bacillus thuringiensis (Bt) that kills certain insects. About 90% of the corn grown in the U.S. was genetically modified in 2010. In the US in 2015, 81% of corn acreage contained the Bt trait and 89% of corn acreage contained the glyphosate-tolerant trait. Corn can be processed into grits, meal and flour as an ingredient in pancakes, muffins, doughnuts, breadings and batters, as well as baby foods, meat products, cereals and some fermented products. Corn-based masa flour and masa dough are used in the production of taco shells, corn chips and tortillas.

Soy

Genetically modified soybean has been modified to tolerate herbicides and produce healthier oils. In 2015, 94% of soybean acreage in the U.S. was genetically modified to be glyphosate-tolerant.

Derivative Products

Corn Starch and Starch Sugars, Including Syrups

Starch or amylum is a polysaccharide produced by all green plants as an energy store. Pure starch is a white, tasteless and odourless powder. It consists of two types of molecules: the linear and helical amylose and the branched amylopectin. Depending on the plant, starch generally contains 20 to 25% amylose and 75 to 80% amylopectin by weight.

Starch can be further modified to create modified starch for specific purposes, including creation of many of the sugars in processed foods. They include:

- Maltodextrin, a lightly hydrolyzed starch product used as a bland-tasting filler and thickener.

- Various glucose syrups, also called corn syrups in the US, viscous solutions used as sweeteners and thickeners in many kinds of processed foods.

- Dextrose, commercial glucose, prepared by the complete hydrolysis of starch.

- High fructose syrup, made by treating dextrose solutions with the enzyme glucose isomerase, until a substantial fraction of the glucose has been converted to fructose.

- Sugar alcohols, such as maltitol, erythritol, sorbitol, mannitol and hydrogenated starch hydrolysate, are sweeteners made by reducing sugars.

Lecithin

Lecithin is a naturally occurring lipid. It can be found in egg yolks and oil-producing plants. it is an emulsifier and thus is used in many foods. Corn, soy and safflower oil are sources of lecithin, though the majority of lecithin commercially available is derived from soy.Sufficiently processed lecithin is often undetectable with standard testing practices. According to the FDA, no evidence shows or suggests hazard to the public when lecithin is used at common levels. Lecithin added to foods amounts to only 2 to 10 percent of the 1 to 5 g of phosphoglycerides consumed daily on average. Nonetheless, consumer concerns about GM food extend to such products. This concern led to policy and regulatory changes in Europe in 2000, when Regulation (EC) 50/2000 was passed which required labelling of food containing additives derived from GMOs, including lecithin. Because of the difficulty of detecting the origin of derivatives like lecithin with current testing practices, European regulations require those who wish to sell lecithin in Europe to employ a comprehensive system of Identity preservation (IP).

Sugar

The US imports 10% of its sugar, while the remaining 90% is extracted from sugar beet and sugarcane. After deregulation in 2005, glyphosate-resistant sugar beet was extensively adopted in the United States. 95% of beet acres in the US were planted with glyphosate-resistant seed in 2011. GM sugar beets are approved for cultivation in the US, Canada and Japan; the vast majority are grown in the US. GM beets are approved for import and consumption in Australia, Canada, Colombia, EU, Japan, Korea, Mexico, New Zealand, Philippines, Russian Federation and Singapore. Pulp from the refining process is used as animal feed. The sugar produced from GM sugarbeets contains no DNA or protein—it is just sucrose that is chemically indistinguishable from sugar produced from non-GM sugarbeets. Independent analyses conducted by internationally recognized laboratories found that sugar from Roundup Ready sugar beets is identical to the sugar from comparably grown conventional (non-Roundup Ready) sugar beets. And, like all sugar, sugar from Roundup Ready sugar beets contains no genetic material or detectable protein (including the protein that provides glyphosate tolerance).

Vegetable Oil

Most vegetable oil used in the US is produced from GM crops canola, corn, cotton and soybeans. Vegetable oil is sold directly to consumers as cooking oil, shortening and margarine and is used in prepared foods. There is a vanishingly small amount of protein or DNA from the original crop in vegetable oil. Vegetable oil is made of triglycerides extracted from plants or seeds and then refined and may be further processed via hydrogenation to turn liquid oils into solids. The refining process removes all, or nearly all non-triglyceride ingredients.

Other Uses

Animal Feed

Livestock and poultry are raised on animal feed, much of which is composed of the leftovers from processing crops, including GM crops. For example, approximately 43% of a canola seed is oil. What remains after oil extraction is a meal that becomes an ingredient in animal feed and contains canola protein. Likewise, the bulk of the soybean crop is grown for oil and meal. The high-protein defatted and toasted soy meal becomes livestock feed and dog food. 98% of the US soybean crop goes for livestock feed. In 2011, 49% of the US maize harvest was used for livestock feed (including the percentage of waste from distillers grains). "Despite methods that are becoming more and more sensitive, tests have not yet been able to establish a difference in the meat, milk, or eggs of animals depending on the type of feed they are fed. It is impossible to tell if an animal was fed GM soy just by looking at the resulting meat, dairy, or egg products. The only way to verify the presence of GMOs in animal feed is to analyze the origin of the feed itself."

A 2012 literature review of studies evaluating the effect of GM feed on the health of animals did not find evidence that animals were adversely affected, although small biological differences were occasionally found. The studies included in the review ranged from 90 days to two years, with several of the longer studies considering reproductive and intergenerational effects.

Proteins

Rennet is a mixture of enzymes used to coagulate milk into cheese. Originally it was available only from the fourth stomach of calves, and was scarce and expensive, or was available from microbial sources, which often produced unpleasant tastes. Genetic engineering made it possible to extract rennet-producing genes from animal stomachs and insert them into bacteria, fungi or yeasts to make them produce chymosin, the key enzyme. The modified microorganism is killed after fermentation. Chymosin is isolated from the fermentation broth, so that the Fermentation-Produced Chymosin (FPC) used by cheese producers has an amino acid sequence that is identical to bovine rennet. The majority of the applied chymosin is retained in the whey. Trace quantities of chymosin may remain in cheese.

FPC was the first artificially produced enzyme to be approved by the US Food and Drug Administration. FPC products have been on the market since 1990 and as of 2015 had yet to be surpassed in commercial markets. In 1999, about 60% of US hard cheese was made with FPC. Its global market share approached 80%. By 2008, approximately 80% to 90% of commercially made cheeses in the US and Britain were made using FPC.

In some countries, recombinant (GM) bovine somatotropin (also called rBST, or bovine growth hormone or BGH) is approved for administration to increase milk production. rBST may be present in milk from rBST treated cows, but it is destroyed in the digestive system and even if directly injected into the human bloodstream, has no observable effect on humans. The FDA, World Health Organization, American Medical Association, American Dietetic Association and the National Institutes of Health have independently stated that dairy products and meat from rBST-treated cows are safe for human consumption. However, on 30 September 2010, the United States Court of Appeals, Sixth Circuit, analyzing submitted evidence, found a "compositional difference" between milk from rBGH-treated cows and milk from untreated cows. The court stated that milk from rB-GH-treated cows has: increased levels of the hormone Insulin-like growth factor 1 (IGF-1); higher fat content and lower protein content when produced at certain points in the cow's lactation cycle; and more somatic cell counts, which may "make the milk turn sour more quickly."

Livestock

Genetically modified livestock are organisms from the group of cattle, sheep, pigs, goats, birds, horses and fish kept for human consumption, whose genetic material (DNA) has been altered using genetic engineering techniques. In some cases, the aim is to introduce a new trait to the animals which does not occur naturally in the species, i.e. transgenesis.

A 2003 review published on behalf of Food Standards Australia New Zealand examined transgenic experimentation on terrestrial livestock species as well as aquatic species such as fish and shellfish. The review examined the molecular techniques used for experimentation as well as techniques for tracing the transgenes in animals and products as well as issues regarding transgene stability.

Some mammals typically used for food production have been modified to produce non-food products, a practice sometimes called Pharming.

Salmon

A GM salmon, awaiting regulatory approval since 1997, was approved for human consumption by the American FDA in November 2015, to be raised in specific land-based hatcheries in Canada and Panama.

Recombinant Food-grade Organisms for Healthcare

The use of genetically modified food-grade organisms as recombinant vaccine expression hosts and delivery vehicles can open new avenues for vaccinology. Considering that oral immunization is a beneficial approach in terms of costs, patient comfort, and protection of mucosal tissues, the use of food-grade organisms can lead to highly advantageous vaccines in terms of costs, easy administration, and safety. The organisms currently used for this purpose are bacteria (Lactobacillus and Bacillus), yeasts, algae, plants, and insect species. Several such organisms are under clinical evaluation, and the current adoption of this technology by the industry indicates a potential to benefit global healthcare systems.

Health and Safety

There is a scientific consensus that currently available food derived from GM crops poses no greater risk to human health than conventional food, but that each GM food needs to be tested on a case-by-case basis before introduction. Nonetheless, members of the public are much less likely than scientists to perceive GM foods as safe.

Opponents claim that long-term health risks have not been adequately assessed and propose various combinations of additional testing, labeling or removal from the market. The advocacy group European Network of Scientists for Social and Environmental Responsibility (ENSSER), disputes the claim that "science" supports the safety of current GM foods, proposing that each GM food must be judged on case-by-case basis. The Canadian Association of Physicians for the Environment called for removing GM foods from the market pending long term health studies. Multiple disputed studies have claimed health effects relating to GM foods or to the pesticides used with them.

Testing

The legal and regulatory status of GM foods varies by country, with some nations banning or restricting them, and others permitting them with widely differing degrees of regulation. Countries such as the United States, Canada, Lebanon and Egypt use *substantial equivalence* to determine if further testing is required, while many countries such as those in the European Union, Brazil and China only authorize GMO cultivation on a case-by-case basis. In the U.S. the FDA determined that GMO's are "Generally Recognized as Safe" (GRAS) and therefore do not require additional testing if the GMO product is substantially equivalent to the non-modified product. If new substances are found, further testing may be required to satisfy concerns over potential toxicity, allergenicity, possible gene transfer to humans or genetic outcrossing to other organisms.

Regulation

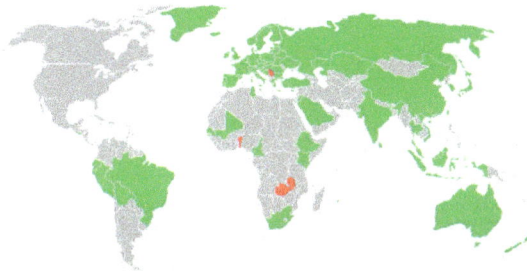

Green: Mandatory labeling required; Red:Ban on import and cultivation of genetically engineered food.

Government regulation of GMO development and release varies widely between countries. Marked differences separate GMO regulation in the U.S. and GMO regulation in the European Union. Regulation also varies depending on the intended product's use. For example, a crop not intended for food use is generally not reviewed by authorities responsible for food safety.

United States Regulations

In the U.S., three government organizations regulate GMOs. The FDA checks the chemical composition of organisms for potential allergens. The United States Department of Agriculture (USDA) supervises field testing and monitors the distribution of GM seeds. The United States Environmental Protection Agency (EPA) is responsible for monitoring pesticide usage, including plants modified to contain proteins toxic to insects. Like USDA, EPA also oversees field testing and the distribution of crops that have had contact with pesticides to ensure environmental safety. In 2015 the Obama administration announced that it would update the way the government regulated GM crops.

In 1992 FDA published "Statement of Policy: Foods derived from New Plant Varieties." This statement is a clarification of FDA's interpretation of the Food, Drug, and Cosmetic Act with respect to foods produced from new plant varieties developed using recombinant deoxyribonucleic acid (rDNA) technology. FDA encouraged developers to consult with the FDA regarding any bioengineered foods in development. The FDA says developers routinely do reach out for consultations. In 1996 FDA updated consultation procedures.

Labeling

As of 2015, 64 countries require labeling of GMO products in the marketplace.

US and Canadian national policy is to require a label only given significant composition differences or documented health impacts, although some individual US states (Vermont, Connecticut and Maine) enacted laws requiring them. In July 2016, Public Law 114-214 was enacted to regulate labeling of GMO food on a national basis.

In some jurisdictions, the labeling requirement depends on the relative quantity of GMO in the product. A study that investigated voluntary labeling in South Africa found that 31% of products labeled as GMO-free had a GM content above 1.0%.

In Europe all food (including processed food) or feed that contains greater than 0.9% GMOs must be labelled.

Detection

Testing on GMOs in food and feed is routinely done using molecular techniques such as PCR and bioinformatics.

In a January 2010 paper, the extraction and detection of DNA along a complete industrial soybean oil processing chain was described to monitor the presence of Roundup Ready (RR) soybean: "The amplification of soybean lectin gene by end-point polymerase chain reaction (PCR) was successfully achieved in all the steps of extraction and refining processes, until the fully refined soybean oil. The amplification of RR soybean by PCR assays using event-specific primers was also achieved for all the extraction and refining steps, except for the intermediate steps of refining (neutralisation, washing and bleaching) possibly due to sample instability. The real-time PCR assays using specific probes confirmed all the results and proved that it is possible to detect and quantify genetically modified organisms in the fully refined soybean oil. To our knowledge, this has never been reported before and represents an important accomplishment regarding the traceability of genetically modified organisms in refined oils."

According to Thomas Redick, detection and prevention of cross-pollination is possible through the suggestions offered by the Farm Service Agency (FSA) and Natural Resources Conservation Service (NRCS). Suggestions include educating farmers on the importance of coexistence, providing farmers with tools and incentives to promote coexistence, conduct research to understand and monitor gene flow, provide assurance of quality and diversity in crops, provide compensation for actual economic losses for farmers.

Controversies

The genetically modified foods controversy consists of a set of disputes over the use of food made from genetically modified crops. The disputes involve consumers, farmers, biotechnology companies, governmental regulators, non-governmental organizations, environmental and political activists and scientists. The major disagreements include whether GM foods can be safely consumed, harm the environment and/or are adequately tested and regulated. The objectivity of scientific research and publications has been challenged. Farming-related disputes include the use and impact of pesticides, seed production and use, side effects on non-GMO crops/farms, and potential control of the GM food supply by seed companies.

The conflicts have continued since GM foods were invented. They have occupied the media, the courts, local, regional and national governments and international organizations.

Reference

- Martins VAP (2008). "Genomic Insights into Oil Biodegradation in Marine Systems". Microbial Biodegradation: Genomics and Molecular Biology. Caister Academic Press. ISBN 978-1-904455-17-2.

- Campbell-Platt, Geoffrey (26 August 2011). Food Science and Technology. Ames, IA: John Wiley & Sons. ISBN 978-1-4443-5782-0.

- Press, Associated (2010-03-03). "GM potato to be grown in Europe". The Guardian. ISSN 0261-3077. Retrieved 2016-02-15.

- Final Report of the PABE research project (December 2001). "Public Perceptions of Agricultural Biotechnologies in Europe". Commission of European Communities. Retrieved February 24, 2016.

- Final Report of the PABE research project (December 2001). "Public Perceptions of Agricultural Biotechnologies in Europe". Commission of European Communities. Retrieved February 24, 2016.

- Chokshi, Niraj (9 May 2014). "Vermont just passed the nation's first GMO food labeling law. Now it prepares to get sued". The Washington Post. Retrieved 19 January 2016.

- Webster, TM; Grey, TL (2015). "Glyphosate-Resistant Palmer Amaranth (Amaranthus palmeri) Morphology, Growth, and Seed Production in Georgia.". Weed Science. 63 (1): 264–272. doi:10.1614/ws-d-14-00051.1.

- Krimsky, Sheldon (2015). "An Illusory Consensus behind GMO Health Assessment" (PDF). Science, Technology, & Human Values. 40: 1–32. doi:10.1177/0162243915598381.

- Cowan, Tadlock (18 Jun 2011). "Agricultural Biotechnology: Background and Recent Issues" (PDF). Congressional Research Service (Library of Congress). pp. 33–38. Retrieved 27 September 2015.

- "Questions & Answers on Food from Genetically Engineered Plants". US Food and Drug Administration. 22 Jun 2015. Retrieved 29 September 2015.

- Webster, TM; Grey, TL (2015). "Glyphosate-Resistant Palmer Amaranth (Amaranthus palmeri) Morphology, Growth, and Seed Production in Georgia.". Weed Science. 63 (1): 264–272. doi:10.1614/ws-d-14-00051.1.

- Rosales-Mendoza, S.; Angulo, C.; Meza, B. (2015). "Food-Grade Organisms as Vaccine Biofactories and Oral Delivery Vehicles". Trends in Biotechnology. doi:10.1016/j.tibtech.2015.11.007.

- Pollack, Andrew (2015-07-02). "White House Orders Review of Rules for Genetically Modified Crops". The New York Times. ISSN 0362-4331. Retrieved 2015-07-03.

Permissions

All chapters in this book are published with permission under the Creative Commons Attribution Share Alike License or equivalent. Every chapter published in this book has been scrutinized by our experts. Their significance has been extensively debated. The topics covered herein carry significant information for a comprehensive understanding. They may even be implemented as practical applications or may be referred to as a beginning point for further studies.

We would like to thank the editorial team for lending their expertise to make the book truly unique. They have played a crucial role in the development of this book. Without their invaluable contributions this book wouldn't have been possible. They have made vital efforts to compile up to date information on the varied aspects of this subject to make this book a valuable addition to the collection of many professionals and students.

This book was conceptualized with the vision of imparting up-to-date and integrated information in this field. To ensure the same, a matchless editorial board was set up. Every individual on the board went through rigorous rounds of assessment to prove their worth. After which they invested a large part of their time researching and compiling the most relevant data for our readers.

The editorial board has been involved in producing this book since its inception. They have spent rigorous hours researching and exploring the diverse topics which have resulted in the successful publishing of this book. They have passed on their knowledge of decades through this book. To expedite this challenging task, the publisher supported the team at every step. A small team of assistant editors was also appointed to further simplify the editing procedure and attain best results for the readers.

Apart from the editorial board, the designing team has also invested a significant amount of their time in understanding the subject and creating the most relevant covers. They scrutinized every image to scout for the most suitable representation of the subject and create an appropriate cover for the book.

The publishing team has been an ardent support to the editorial, designing and production team. Their endless efforts to recruit the best for this project, has resulted in the accomplishment of this book. They are a veteran in the field of academics and their pool of knowledge is as vast as their experience in printing. Their expertise and guidance has proved useful at every step. Their uncompromising quality standards have made this book an exceptional effort. Their encouragement from time to time has been an inspiration for everyone.

The publisher and the editorial board hope that this book will prove to be a valuable piece of knowledge for students, practitioners and scholars across the globe.

Index